BIOLOGICAL NOMENCI

A review of the present state and current issues of biological nomenclature of animals, plants, bacteria and viruses

Papers presented at ICSEB III
July 9, 1985
Brighton, United Kingdom

Edited by

Dr. W.D.L.Ride
School of Applied Science
Canberra College of Advanced Education
POB 1
Canberra ACT 2616, Australia

and

Dr.T.Younès
International Union of Biological Sciences
51 Boulevard de Montmorency
75016 Paris, France

Published by IRL Press Limited
PO Box 1 Eynsham
Oxford OX8 1JJ, UK

on behalf of the

ICSU Press
PO Box 016129
Miami, Florida 33101, USA

for the

International Union of Biological Sciences
51 Boulevard de Montmorency
75016 Paris, France

First published 1986

ISBN 1-85221-016-8

IUBS Monograph Series, No. 2

Biological Nomenclature Today is the second in the IUBS Monograph Series. Other volumes are:

No. 1. Biological Monitoring of the State of the Environment: Bioindicators

No. 3. Determinants of Tropical Savannas

No. 4. New Challenges for Biological Education

Monographs 1 and 3 are available from IRL Press.
Monograph no. 4 is available from the IUBS Secretariat at a cost of $5 post paid.

THE INTERNATIONAL UNION OF BIOLOGICAL SCIENCES

The International Union of Biological Sciences is a non-governmental, non-profit organization, established in 1919. Its objectives are to promote the study of biological sciences, to initiate, facilitate, and coordinate research and other scientific activities that require international cooperation, to ensure the discussion and dissemination of the results of cooperative research, to promote the organization of international conferences and to assist in the publication of their reports.

The membership of IUBS presently consists of 44 Ordinary Members, adhering through Academies of Science, National Research Councils, national science associations or similar organizations, and of 69 scientific members, all of which are international scientific associations, societies or commissions in the various biological disciplines.

The IUBS publishes Biology International, a news magazine, currently appearing as two regular issues and two special issues per annum, for which the subscription price is $20. It is published with the financial support of Unesco's 'Man and the Biosphere' programme.

The Editorial Board consists of F. di Castri (France), P. Fasella (Italy), W.D.L. Ride (Australia), E. de Robertis (Argentina), D.F. Roberts (UK), J. Salanki (Hungary), V.E. Sokolov (USSR), O.T. Solbrig (USA), and T. Younès (IUBS Secretariat, Editor).

Further information on IUBS and Biology International may be obtained from its Executive Secretary, Dr T.Younès, at 51 Boulevard de Montmorency, 75016 Paris, France.

THIS IS ICSU

This monograph is published for the International Union of Biological Sciences (IUBS) by the ICSU Press, in conjunction with IRL Press.

The ICSU Press is the publishing house of the International Council of Scientific Unions, an international non-governmental scientific organization, whose principal objective is to encourage international scientific activity for the benefit of all. ICSU's membership is composed of 20 Scientific Unions, representing various disciplines and national bodies such as academies or research councils in approximately 70 countries. The IUBS is one of the eight biological Unions adhering to ICSU.

Since its creation in 1931 ICSU has adopted a policy of non-discrimination, affirming the rights of all scientists throughout the world — without regard to race, religion, political philosophy, ethnic origin, citizenship, sex or language — to join in international scientific activities.

To fulfil its objectives, ICSU initiates, designs and coordinates international interdisciplinary research programmes, such as the project for the study of the interactions between various parts of the biosphere and geosphere, known as the Global Change programme, which was launched at the 21st General Assembly in 1986.

ICSU acts as a focus for the exchange of ideas, the communication of scientific information, and the development of scientific standards, nomenclature, units, etc.

Members of the ICSU family organize in many parts of the world scientific conferences, congresses, symposia, summer schools and meetings of experts, as well as General Assemblies and other meetings to decide policies and programmes. In 1985 more than 500 meetings were organized.

ICSU members produce a wide range of publications, including newsletters, handbooks, proceedings of meetings, congresses and symposia, professional scientific journals, data, standards, etc. Some of these are published by the ICSU Press, including BioEssays, the monthly current-awareness journal, sponsored by ICSU's biological Unions.

For programmes in multi- or transdisciplinary fields, such as Antarctic, Oceanic, Space Research, or Genetic Experimentation, which are not under the aegis of one of the Scientific Unions, and for activities in areas common to all the Unions, such as Teaching of Science, Data, Science and Technology in Developing Countries, ICSU creates Scientific Committees or Commissions.

ICSU maintains close relations and works in cooperation with a number of international governmental and non-governmental organizations.

Further information about ICSU and the ICSU Press may be obtained from the ICSU Secretariat, 51 Boulevard de Montmorency, 75016 Paris, France.

CONTENTS

Acknowledgement

The Editors would like to express their deep appreciation to the contributors of this publication, who have offered their advice, comments and suggestions, as well as to The Systematics Association. They also wish to thank Mrs. C. Adam for her valuable assistance in the preparation of this report.

INTRODUCTION

Biological nomenclature is fundamentally important to international scientific communication. By adopting its standards for self-regulation in nomenclature, all biological scientists, whether they be bacteriologists, botanists, virologists or zoologists, can arrive at and use the same name for the same taxon. Moreover they can do so without restricting in any way the taxonomy upon which biological science depends to define the organisms that it studies.

The bodies responsible for administering and updating the botanical and zoological codes of nomenclature are scientific members of the International Union of Biological Sciences. It is unthinkable that biological nomenclature, and its support and well-being, should be considered as other than a matter of international concern and part of the continuing Scientific Programme of IUBS.

The papers presented in this volume were developed by the authors from those presented at an International Congress of Systematic and Evolutionary Biology (ICSEB III) held on 9 July, 1985, at the University of Sussex, Brighton, England. Papers given at two sessions of that Congress are represented, namely a session entitled "Codes of Nomenclature" organized by R.V. Melville on behalf of the Systematics Association, and the other, a workshop, convened by W.D.L. Ride on behalf of the Executive of IUBS to study future developments of the various codes of nomenclature and, in particular, issues of importance to the treatment of organisms (such as protists) currently under more than one code. The latter followed a session entitled "The Protists: Evolution, Taxonomy and Nomenclature", convened by Dr. J.O. Corliss.

Participants at both sessions found that the opportunity to discuss issues in nomenclature common to all biologists in open forum was so useful that they requested ICSEB to provide regular opportunities at its Congresses for similar sessions in future. Moreover, the International Commission on Zoological Nomenclature subsequently sought concurrence of the International Union of Biological Sciences (through the Executive) to authorize its Section of Zoological Nomenclature to hold meetings at ICSEB Congresses. (The Section of Nomenclature of IUBS is required to elect members of the Commission and approve proposed amendments to the zoological code.) The International Commission on Zoological Nomenclature hopes that meetings held at ICSEB (and possibly other Congresses as well) will fill the gap in communication with systematic zoologists that had resulted from the demise of the former International Congresses of Zoology. Both ICSEB and IUBS have agreed, and it is confidently expected that a new and fruitful association will develop that will, in addition, bring closer together botanical and zoological systems (together with those responsible for the nomenclature of bacteria and viruses[1]).

----

[1] Currently with the International Union of Microbiological Societies (IUMS).

2

The papers included here are published as a contribution of the Scientific Programme of IUBS, and in the hope of creating a better understanding of the role of the four nomenclatural systems within the member bodies of IUBS and of the scientific community generally.

Together with the taxonomy it serves, nomenclature is basic to biology. The Scientific Programme of IUBS is committed to the aims of both, which are to discover, define, name and understand the relationships and distribution in time and space of the world's organisms.

BOTANICAL NOMENCLATURE

J. McNeill
and
W. Greuter[1]

This review of botanical nomenclature today owes its origin to two papers presented on 9 July 1985 at sessions of the Third International Congress of Systematic and Evolutionary Biology (ICSEB III) held at the University of Sussex, Brighton, England. The first, by McNeill, entitled "Botanical nomenclature", was one of six contributions in a morning symposium organized by R.V. Melville, on behalf of the Systematics Association, on "Codes of Nomenclature"; the second, by Greuter, on the history and present organization of botanical nomenclature was presented that afternoon in a workshop, organized by W.D.L. Ride, to discuss the future development of the various codes of nomenclature, considering, in particular, the treatment of groups of organisms coming under more than one code. Together these papers provide an outline of the origins of the present *International Code of Botanical Nomenclature* (ICBN), of the way in which botanical nomenclature functions today, of the major differences between the botanical and other codes of nomenclature, of the philosophical differences and consequent tensions that are inherent in nomenclatural codes, of the current issues facing botanical nomenclature, and of the future directions that botanical nomenclature may take.

The importance of nomenclature to biological communication should first of all be stressed. As van Steenis (1957) put it, "a plant's name is the key to its literature". Without a stable and universally acceptable set of scientific names for plants and animals, unambiguous scientific communication becomes impossible in almost any branch of biology: a molecular biologist needs to be precise about the source of a DNA sequence, as much as a physiologist or ecologist must refer unmistakably to the organisms which he or she studies. Stable and universally acceptable scientific names require internationally agreed rules. The codes of nomenclature are thus in a way basic to almost all scientific endeavours in biology.

The *International Code of Botanical Nomenclature* (ICBN) "applies to names of taxonomic groups treated as plants" (Voss *et al.* 1983, Principle I), and thus avoids making any taxonomic prejudgements as a requirement for its application. In practice, the Code covers all groups of fungi and algae as well

---

[1] Professor Werner Greuter is Rapporteur-général for the XIV International Botanical Congress to be held in Berlin (West), Germany in July 1987; Professor John McNeill is Vice-rapporteur.

as the "green plants" (the "Embryophyta"). The current Botanical Code (Voss *et al.* 1983), revised at the XIII International Botanical Congress (IBC) in Sydney, Australia in 1981, is a fairly substantial work of 472 pages, although 226 pages of these are Appendices and Indices. The main text is in three languages (English, French and German). Although based on only six Principles, the present Code comprises 73 separate Articles and 60 Recommendations. Within the space available here, we must, therefore, be very selective and do no more than address issues of current importance, including the differences between the Botanical and Zoological Codes in particular, and the historical and organizational factors that have caused these differences.

The present Code can be considered to be the thirteenth edition of an internationally adopted set of rules governing the nomenclature of plants, although only the last seven editions (from the "Stockholm Code" of 1952 - Lanjouw *et al.* 1952) have been entitled *International Code of Botanical Nomenclature.* (See Lanjouw *et al.* 1966, pp. 400-402, for full bibliographic details of the 9 "editions" up to that date, and Parkinson 1975, for extensive bibliographic information on codes of nomenclature).

History of Botanical Nomenclature

The origins of a botanical code go back to Alphonse de Candolle's *Lois* in the middle of the last century (de Candolle 1867). These were drafted by de Candolle, reviewed by a commission of 6 European botanists, and discussed at and approved by sessions of a "Congrès International de Botanique tenu à Paris en août 1867 sur les auspices de la Société botanique de France" (Fournier 1867). Although dealing with many issues that today would be considered advisory rather than mandatory (e.g. that botanists avoid "les noms très longs ou d'une prononciation difficile"), several contemporary nomenclatural principles were established, including priority of publication and that each taxonomic group should bear only one correct name. Although priority was established as the criterion for choice between competing names, several important aspects of priority were not considered, leading to later confusion or conflict. Specifically, the matter of priority upon transfer, e.g. when a species is transferred to a different genus, was not addressed. This led to what might be called the first "schism": the development of the "Kew Rule", in which combinations as such, and not epithets published as part of a combination, had priority. Thus a specific epithet would not necessarily have priority outside the genus in which it was published. This rule (which was clearly rejected at the Vienna Congress of 1905) was adopted in the choice of correct names in the main volumes and first supplements of *Index Kewensis,* as well as in other British and some American publications of the latter part of the 19th century. (The American publications were primarily of those who followed the Harvard tradition of Gray and Robinson).

Moreover, although the name adopted by Linnaeus was considered to be the oldest, de Candolle did not at first recognize the problems inherent in Linnaeus's own change of usage of generic names between the first edition of *Genera Plantarum* in 1737 and that of *Species Plantarum* in 1753 (but see de

Candolle 1883). Otto Kuntze (Kuntze 1891-98) did, however, and released a flood of unfamiliar generic names, which he considered had precedence but which other plant taxonomists were unwilling to adopt. These issues prompted the Paris IBC of 1900 to decide that a revision of the *Lois* was required. A Commission - of 46 members, of whom 4 were from the U.S. - was established to report to the Vienna Congress of 1905. Meanwhile, and in preparation for that Congress, the Botanical Club of the American Association for the Advancement of Science (AAAS) established its own Nomenclatural Commission, which concluded that a completely new code was required, and published a proposed text - interestingly in French and German as well as in English (Arthur *et al.* 1904). This text was developed from the so-called "Rochester Code", which was little more than an enunciation of principles (Anonymous 1892; Fairchild 1892). (For a useful bibliography of the debates prior to the Vienna Congress, both among European botanists and between the "Rochester / New York" and "Harvard" schools of U.S. botanists, see Parkinson 1975).

After debates over six sessions, the Vienna Congress adopted a revision of de Candolle's *Lois*, rather than accept the draft code of the Botanical Club of the AAAS (see Rendle (1905) and Cheeseman (1908) for contemporary accounts, and Smith (1957) for an historical perspective). This revision was given the title *Règles internationales de la Nomenclature botanique* (Briquet 1906) and is often called the "First edition of the Rules", or, more often nowadays, just as the "Vienna Code". In addition to prescribing absolute priority within a rank - thus already 80 years ago establishing an important difference from the zoological code's practice of adopting coordinate status[2] between different ranks in the "species-group" and "genus-group" - the Vienna Congress responded to the unwelcome work of Otto Kuntze by approving an extensive list of over 400 names in an *Index nominum genericorum utique conservandorum*. The transatlantic response was quick: an opposition "American Code" was published - and only in English (Arthur *et al.* 1907). This accepted priority within ranks (the 1904 draft had envisaged something like coordinate status at the specific and subspecific levels), but the Americans were disappointed that the Vienna Congress had failed to recognize the concept of nomenclatural types and were convinced, rightly as it transpired, "that the method by types will obtain general recognition and acceptance, inasmuch as it is the only one which promises sufficient definiteness". They were, moreover, very critical of the provision for conservation of generic names, an action which they viewed "as in the

----

[2] Coordinate status refers to the "Principle of Coordination" in the *International Code of Zoological Nomenclature* (ICZN) (see Melville, this volume, p.   ), in which a name, published for a taxon at any rank of a particular "group" of ranks, is deemed to be simultaneously established with the same author, date and type, at all other ranks in that group. The ICZN has three such groups: the "family-group" (all ranks above genus up to superfamily); the "genus-group" (genus and subgenus); and the "species-group" (species and subspecies). Cf. Weresub (1970) and Brummitt (1981) for discussion of coordinate status in relation to the Botanical Code.

highest degree arbitrary" and "as controverting a cardinal principle"
(i.e. priority). It is perhaps noteworthy that, even today, opposition to
proposals for conservation of names seems more likely to come from
U.S. members of plant nomenclature committees than from European ones, and
that it seemed largely to be the numerous proxied U.S. institutional votes
which, in Leningrad in 1975, defeated the proposal for unqualified
conservation of species' names, a proposal that had enjoyed the support of a
substantial majority of the individuals present.

By the end of World War I (1914-18), the need to reconcile the
International and American Codes was apparent on both sides of the Atlantic.
Isolationism may have dominated U.S. politics at that time, but it was
overcome in U.S. botany, largely at the instigation of A.S. Hitchcock, who put
forward a "Type-basis code" (Hitchcock 1919, 1921), as a proposal for
international acceptance and not as a complete "take-it-or-leave-it"
nomenclatural code (see also Hitchcock 1925, ch. 15). The first Botanical
Congress held outside of Europe - the Ithaca Congress of 1925 - paved the
way for the rapprochement that was achieved at the Fifth Congress at
Cambridge, England in 1930. These "Cambridge Rules", or the "Third edition
of the Rules" (Briquet 1935), incorporated the method of determining the
application of names by means of nomenclatural types, the contribution of the
American Code, but dropped the detailed provisions for the determination of
types that had been a feature of that Code. The Cambridge Rules retained
nomina generica conservanda but, in deference to American tradition, deferred
until 1 January 1935, the requirement for a Latin diagnosis for valid
publication of new names which had been made mandatory in the Vienna
Code from 1st January 1908.

No code embodying the further amendments made at the Sixth Congress
in Amsterdam in 1935, appeared until after World War II. Largely through
the influence of Lanjouw, a nomenclatural conference was organized in 1948
with the support of the International Union of Biological Sciences (IUBS).
This allowed effective preparations to be made for the first post-war
International Botanical Congress, held in Stockholm, in neutral Sweden, in
1950. At the same time, and also due to Lanjouw's initiatives, a secure
operational and financial basis for botanical nomenclature was established
through the formation of the International Association for Plant Taxonomy
(IAPT) (Lanjouw 1950; Cowan and Stafleu 1982; Stafleu 1985).

The 45 years from the publication of the Vienna Code to the Stockholm
Congress, had seen only three revisions to the International Rules (at Brussels,
Cambridge, and Amsterdam; see Lanjouw *et al.* 1966, pp. 400-402). The
Stockholm Code was published in 1952 (Lanjouw *et al.* 1952) and since then
a revised code has been published following each International Botanical
Congress. The VIII Congress was held in Paris in 1954. From then until
1969, Congresses were held every 5 years; since then they have been held
every 6 years, so that 7 editions of the International Code appeared between
1952 and 1983. Although important revisions to the Articles and
Recommendations have been made at nearly every Congress (e.g. the addition
of a list of *Nomina familiarum conservanda* at Montréal in 1959 (Lanjouw *et
al.* 1961), and the provision for conservation of names of species of major

economic importance at Sydney in 1981 (Voss *et al.* 1983)), no new basic concepts have been developed. It has indeed been suggested on some occasions, notably at the "conservative" Leningrad Congress of 1975 (cf. Voss 1979; pp. 174-175), that no new edition of the Code be published, but in fact this has never been acted upon.

International rules of botanical nomenclature have thus had a long history (for other useful accounts, see Lawrence 1951, ch. 9, Smith 1957, and Parkinson 1975). The codes of nomenclature have, for the most part, been followed faithfully by botanists for over 100 years. Any amendment to the Code almost certainly requires some names, that were correct under previous versions, to be changed. This means that amendments that will not demonstrably improve nomenclatural stability (e.g. by clarifying wording that has been shown to be ambiguous) are unlikely to be accepted. Likewise, the potential benefit of amendments that will reduce discrepancies between the different codes (i.e. botanical, zoological, and bacteriological), must be weighed against the extent to which these same amendments will cause presently correct names to be made incorrect.

Organization of Botanical Nomenclature

The basic administrative structure of botanical nomenclature is summarized in Division III of the Botanical Code - "Provisions for the Modification of the Code" (Voss *et al.* 1983, pp. 70-71; see also McVaugh et al. 1968 pp. 28-30, and Greuter and McNeill 1983). Essentially, this specifies that the Code may be modified only by the plenary session of an International Botanical Congress acting on the basis of proposals approved by the Nomenclature Section of the Congress, which meets prior to the main Congress sessions and at which any botanist attending the Congress may register, but which in practice is attended by about 150 plant taxonomists with particular interests in nomenclature, each of whom has the right to exercise a single vote. In addition, taxonomic institutions have votes in the Nomenclature Section, assigned on a scale of 1 to 7, and these institutional votes, unlike individual votes, may be exercised by proxies; however no individual may exercise more than 15 votes, his personal vote included. As a result, the Nomenclature Section at the Sydney Congress had a total of about 425 registered votes. In all recent Congresses, a 60% majority of the votes cast has been required for approval by the Nomenclature Section of any proposal to amend the Code.

The organization of the Nomenclature Section is in the hands of the Bureau of Nomenclature of the Congress, comprising a Rapporteur-général, elected by the previous Congress, a President and a Recorder appointed by the Congress ·organizing committee, and a Vice-rapporteur appointed by the organizing committee on the proposal of the rapporteur-général.

The Nomenclature Section of an IBC is, however, merely the final stage in the process of amending the Code, and in itself this represents only a part of the ongoing activities of botanical nomenclature (see Table 1). Proposals to amend the Code are published in *Taxon*, the journal of IAPT (see p. 17), are reviewed by the rapporteurs (rapporteur-général and vice-rapporteur), whose comments are also published in *Taxon* as part of a "Synopsis of Proposals" (see Greuter and McNeill (1983) for an explanation of the procedure, and Voss and Greuter (1981) for the most recent Synopsis).

Nomenclatural activity between Congresses is carried out by the General Committee (which does not exist while the Nomenclature Section is in session, but which represents it between Congresses), an Editorial Committee for the Code, 7 other "Permanent Nomenclatural Committees" dealing with particular groups covered by the Code (e.g. Fungi and Lichens, Spermatophyta, Hybrids, etc. - see Table 1), a number of *ad hoc* committees (referred to here as "Special Committees")[3] set up to report to the next Congress. The General Committee on Botanical Nomenclature, in addition to at least 5 (currently 8) members at large, is made up of the Secretaries of the other Permanent Committees, the Rapporteur-général, and the President and Secretary of IAPT. Table 1 illustrates the organizational structure of plant nomenclature.

The General Committee is the body primarily responsible for botanical nomenclature between Botanical Congresses. It makes preliminary decisions on recommendations for conservation and rejection of names made to it by the Permanent Committees for special groups; it takes action to determine whether or not names are sufficiently alike to be confused and makes recommendations on such cases for binding decision by the next Congress; it appoints the members of any Special Committees set up by the Nomenclature Section to study and report on controversial, confused or complicated issues; it generally advises the Bureau of Nomenclature of the succeeding Congress on an appropriate allocation of institutional votes.

The chief work of the Permanent Committees for special groups is to consider proposals to override the provisions of the Code in specific cases through conservation or rejection of names. Almost 100 plant nomenclaturalists are involved, through these various committees, in the ongoing work of plant nomenclature. The work load of the Secretaries of some of these Permanent Committees is substantial and several have this recognized as part of their formal duties in the Institutes in which they are employed. W.D.L. Ride, in his report to IUBS on the method of operation of the different codes of nomenclature (Ride 1984), gives a clear account of this decentralization of botanical nomenclature in contrast to the situation in zoological nomenclature (but see Ride, this volume.).

---

[3] The terminology used for the various plant nomenclature committees is not always uniform. The Botanical Code (Division III. Provisions for Modification of the Code) specifies the 9 Permanent Nomenclatural Committees referred to above. The 7 Permanent Committees for particular groups or for "special groups" are sometimes also called "special committees", but technically these are the "Permanent Committee for . . . " ("Spermatophyta" etc.). Where it is necessary to distinguish them as a group, we shall refer to them as the "Permanent Committees for special groups".

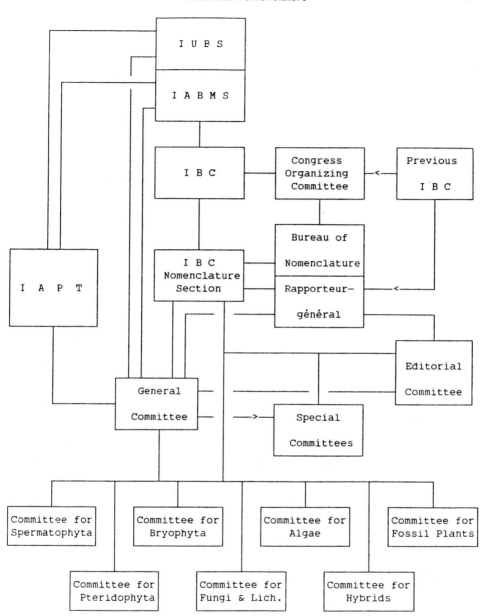

Table 1. Organizational Chart of Botanical Nomenclature

(for key to abbreviations see text and Appendix)

The authority for the Botanical Code thus rests with the successive International Botanical Congresses. The final plenary session of each Congress adopts the proposals of the Nomenclature Section to amend the Code, approves the names proposed for conservation or rejection, and appoints the Rapporteur-général for the next Congress and the members of the various nomenclatural committees.

Recent Congresses have been organized and have met under the aegis of what was then the Division of Botany of IUBS. With the decision to dispense with a the divisional structure for IUBS taken at the XXI General Assembly in Ottawa in 1982 (IUBS 1983), the International Association of Botanical and Mycological Societies (IABMS)[4] was established to take over the role of the former Division of Botany of organizing Botanical Congresses. The IUBS Section of General Botany plays a pivotal role in the IABMS and has done so in the organization of the XIV IBC in Berlin in 1987.

The International Association for Plant Taxonomy (IAPT) plays a vitally important role in plant nomenclature. A special section of its journal, *Taxon*, is the normal medium for publication of nomenclatural matters, e.g. proposals for conservation of names, reports of the standing committees on nomenclature, proposals to amend the Code, the synopses of these proposals, and the nomenclatural decisions of the International Botanical Congresses. The membership of IAPT includes a much wider range of plant taxonomists than those with special interests in nomenclature, and this is reflected in the scope of the papers published in *Taxon*. All members of IAPT are, however, informed, if only through *Taxon*, of all major matters concerning plant nomenclature, and are entitled to vote in the mail ballot on proposals to amend the Code, which is held as a preliminary to each Congress and which is advisory to the Nomenclature Section of the Congress. IAPT forms the Section of Plant Taxonomy of IUBS, whereas the much smaller and more specifically nomenclatural body, the General Committee on Botanical Nomenclature (see above), makes up the IUBS Commission on the Nomenclature of Plants.

---

[4] The International Association of Botanical and Mycological Societies (IABMS) comprises the following 18 organizations, all of which are Sections and Commissions of the IUBS: Commission on the Nomenclature of Cultivated Plants, Eriksson Prize Fund, General Committee on Botanical Nomenclature, International Association of Botanical Gardens, International Association for Plant Taxonomy (IAPT), International Commission on Bee Botany, International Mycological Association, International Phycological Society, International Organization of Palaeobotany, International Organization for Succulent Plant Research International Plant Growth Substances Association, International Seed Testing Association, International Society for Horticultural Science, International Society for Mushroom Science, International Society for Plant Pathology, International Society for Plant Physiology, Plant Protection Congresses, and Section of General Botany.

[Lest confusion arise, a word should be said at this point about a different code - the *International Code of Nomenclature of Cultivated Plants* (ICNCP). Successive editions of this code are formulated and adopted by the IUBS Commission for the Nomenclature of Cultivated Plants. The Commission was initially appointed by IUBS, primarily at the instigation of successive International Horticultural Congresses, but with representatives of agriculture and forestry. It is now self-perpetuating. The history of codes of nomenclature of cultivated plants is outlined in the preface to the most recent code (Brickell *et al.* 1980). As is made clear in the ICNCP, it is in no way in conflict with the Botanical Code because it is concerned only with "the naming of agricultural, horticultural, and silvicultural *cultivars* (*varieties*)" (ICNCP Art. 3) - our emphasis -, whereas "the *International Code of Botanical Nomenclature* . . . governs the use of botanical names in Latin form for both cultivated and wild plants" (ICNCP Art. 2). Thus the ICNCP deals, in a sense, with a lower taxonomic level than the Botanical Code, and, although quite independent of it, provides a subset of rules for cultivars within the nomenclature established by the Botanical Code.]

This organizational structure of botanical nomenclature, with its strong decentralization, has a number of significant results. First of all, it has, up to now at least, enabled botanical nomenclature to operate without major financial problems. There are two aspects to this. On the one hand, the broadly spread workload has allowed the costs of professional activity either to be similarly spread across a fairly large number of botanical institutes, and hence to be accommodated within their budgets or by modest research grants, or else to be voluntarily provided in their own time as a labour of love by the taxonomists concerned. Equally important has been the existence of IAPT, with over 1800 individual and institutional members providing an economic base for the publication of nomenclatural proposals and other essential items in its journal *Taxon*. This broad organizational base of nomenclature and the existence of IAPT have also undoubtedly increased the sales potential of publications such as the ICBN, and other publications in the *Regnum Vegetabile* series.

A second important result of the broad organizational base of botanical nomenclature is that in recent years there have been almost no problems in the acceptance of the Botanical Code throughout the plant taxonomic community. Equally, of course, amendments to the Code are only possible if they command the support of a substantial majority of plant taxonomists. The long slow road towards acceptance of *nomina specifica conservanda* from Montréal in 1959 (Little 1957; Lanjouw 1959, pp. 21-23; Bureau of Nomenclature 1960, pp. 56-68) to Sydney in 1981 (Greuter and McNeill 1981, Voss and Greuter 1981 pp. 106-107), where even then the acceptance was only for species of major economic importance (Greuter and Voss 1982 pp.52-56), illustrates well the necessity for a strong and "democratic" consensus before changes can be made in the botanical code.

The International Code of Botanical Nomenclature and other Codes

Other contributors to this symposium have outlined something of the characteristics of the other major codes of biological nomenclature - notably the *International Code of Zoological Nomenclature* (ICZN) (Ride *et al.* 1985) and the *International Code of Nomenclature of Bacteria* (ICNB) (Lapage *et al.* 1975). Moreover, Voss (1969) has provided an excellent summary, and Jeffrey (1977) a comparative study of these three codes based on the editions that were then current.

Differences between the codes of nomenclature are troublesome for three main reasons: biology students may be confused by the lack of uniformity in terminology (e.g. in name endings); serious confusion can arise, particularly in information retrieval, when the same name is used for two disparate organisms that are the subject of common study (e.g. because they occupy the same ecological area); and, perhaps most seriously, when different names are correct for the same organism, depending on the classification adopted by the taxonomist and hence the Code under which the organism's nomenclature is to be regulated. It is relevant, therefore, to consider here some of the more noticeable or more significant differences between the Botanical Code and these other two codes. As this paper was just presented at a congress of systematic and evolutionary biology, it is perhaps appropriate to use evolutionary idioms to describe the relationships of the Codes.

The origin of the Bacteriological Code in 1958 (Buchanan *et al.* 1958; but see also Buchanan *et al.* 1968) can be looked on as a "speciation event" within the lineage of the successive Botanical Codes, which had applied also to the bacteria. Consequently much of the basic terminology is common to the two Codes, and apart from the distinctive features of a new starting date (1 January 1980 instead of 1 May 1753) along with the requirement for acceptance of earlier names of inclusion in one of the "Approved Lists of Bacterial Names", the acceptance of descriptions in any language, and the lack of any restrictions on conservation or rejection of names, there are relatively few differences between the Codes. Nevertheless, some differences were established right from the beginning and others have developed with amendments to the Botanical Code. For example, the Bacteriological Code, although not explicitly adopting the Principle of Coordination of the Zoological Code, retains the priority of a subspecific epithet when the rank is changed to that of species (and *vice versa*), whereas the Botanical Code provides for priority strictly within each rank. Moreover, the Bacteriological Code adopts the former botanical practice of designating subordinate taxa as the types of the names of taxa above the rank of species (e.g. a species as the type of a generic name). Recognizing that a species is not a single "element", but has a circumscription that in some cases will be controversial, the Botanical Code as amended at Sydney in 1981, requires that all types be specimens (or other elements acceptable as the types of species names); in the case of a generic name, this will normally be the type of the name of the previously designated "type species". The *Pseudolarix* example, discussed under the "pragmatic" and "idealistic" nomenclatural philosophies, below, illustrates this important distinction.

The relationship between the Botanical and Zoological Codes, on the other hand, is much more one of two genetically related populations that have long been allopatric and that have responded in different ways to similar or different selection pressures and have probably also been subject to more than a little genetic drift! Consequently, despite sharing the fundamental principles of priority of publication for selecting between competing names, and of determining the application of names by means of nomenclatural types, the two Codes are very different documents. Moreover, not only have botanists and zoologists often taken different directions in dealing with similar issues, they have also adopted separate and often confusingly different terminologies, for the same or very similar concepts.

There are many examples of this. A few of these are:

Botanical Code	Zoological Code
later homonym	junior homonym
nomenclatural synonym	objective synonym
taxonomic synonym	subjective synonym
validly published	available
correct name	valid name
specific epithet	specific name
specific name	binomen, name of a species

More troublesome, however, to the general user of biological names are the differences in name endings, and in the way in which authors of names are cited. The name ending -*idae* implies a subclass in botany but a family in zoology; the name of a botanical family must end in -*aceae* (with 9 permitted exceptions sanctioned by long usage) and a subfamily in -*oideae* - but this designates a superfamily in zoology; moreover the zoological subfamily ending -*inae* applies to a subtribe in botany. The botanical procedure of the "double citation" of authors' names, in cases in which an epithet has been transferred to another genus or where there has been a change of rank, emphasises the information retrieval function of citation, including not only the source of the type, but also the work in which the currently accepted taxonomy was first established. By contrast the zoological practice of citing only the original author restricts itself to type source identification, irrespective of how inappropriate the taxonomy or rank adopted by the original proposer might have been.

But these are all relatively superficial matters, and compromise would probably not be impossible. Other, more substantial, differences exist, as, for example, the prohibition of tautonyms and the requirement of a Latin diagnosis for valid publication of names of new taxa in the Botanical Code, and the absence of any equivalence in the Zoological Code to the Botanical Code's concept of illegitimacy. These and other differences impinge on the more basic issue, already noted in the discussion of the history of botanical nomenclature, namely that any change in the Code, however small, almost inevitably leads to name changes. A 60% majority of the 400-500 votes at an IBC Nomenclature Section meeting must be convinced that any particular step to achieve greater uniformity between codes, is worth the name changes that it may cause. It is noteworthy that even such an apparently innocuous proposal as that to change the name of the first botanical rank below *regnum* (Kingdom) from *divisio* (Division) to *phylum* (Phylum) (Bold *et al.* 1978), made, for pedagological reasons, to bring the Botanical Code into line with zoological practice (nowhere mentioned, far less legislated, in the Zoological Code, incidentally), failed to gain the necessary 60% majority at Sydney in 1981 (Greuter and Voss 1982 pp. 16-17). This was largely because the proposers had not appreciated that the two terms had been applied, historically, to two different ranks, and in their proposal had not indicated how names published in works using both ranks should be treated.

It is certainly clear that any serious attempt to reconcile the differences between the Botanical and Zoological Codes, requires careful preparation and a recognition of all the nomenclatural implications on both sides. For many of the discrepancies between the Codes, this is probably an impossible dream. The different allopatric populations have diverged too far. To consider why this has happened, why the "genetic makeup" of botanical nomenclature as well as its "selection pressures", may have been different from those of zoological nomenclature, it is necessary to look more fundamentally at approaches to biological nomenclature and to consider some of the ongoing tensions that are an inevitable part of it.

Nomenclatural Philosophies - Tensions in Biological Nomenclature

Some biologists who are not involved in nomenclatural matters tend to believe that nomenclature has become a much too erudite matter and perhaps also too time-consuming and expensive an operation for the stability it seeks to achieve. Why is there this perception? Why, indeed, have codes of nomenclature become lengthy and relatively complicated? Could we not just stop at priority of publication and application of names by means of nomenclatural types? To answer these questions we have to remember the way in which the codes have developed, as, in the case of the Botanical Code, from the "Napoleonic" style of de Candolle's *Lois* with their general rules and user interpretation, through decades in which they have been overlaid by Anglo-Saxon jurisdictional procedures. Within biological nomenclature, as in any other legal system, there are not only historical factors, but also a number of opposing tendencies, and the codes have been developed in an attempt to reconcile these conflicting approaches or philosophies.

One fundamental difficulty that must always be addressed is the maintainance of a consistency between the standards of the past and the requirements of the future. For example, the codes nowadays utilize the type method for determining the application of names, but an 18th or 19th century name should not be disallowed just because its author did not cite a type, a current requirement for valid publication under the Botanical Code. More generally, it is an important element of stability to try to ensure that the nomenclatural actions of authors, who faithfully followed the rules and procedures that were accepted in their day, are not made invalid, unnecessarily, by changes in the rules of nomenclature, introduced to meet the perceived needs of today. This is particularly the case with changes designed to ensure greater precision in the rules.

Another tension in nomenclature is between maintaining a simple code and developing a complex one. There are those who want a relatively simple code with little more than the basic principles, and who, in consequence, are prepared to accept very extensive lists of official rulings or of exceptions to the Code, whenever application of these basic rules is ambiguous or leads to results that are considered to be undesirable. The opposite position is to eschew exceptions and official rulings and to seek to make the rules themselves cover most eventualities even at the expense of developing a relatively involved code.

It is conventional wisdom among botanists that the Zoological Code has tended to favour the first alternative and the Botanical Code the second, but a glance at the most recent issue of the former (Ride *et al.* 1985) makes it clear that even with the greater willingness of zoologists to authorize exceptions (through conservation and rejection of names), and to make official rulings, which scarcely exist in botanical nomenclature (the one case being to determine "whether names are sufficiently alike to be confused"), the Zoological Code is not exactly a simple document! The 88 Articles, the Recommendations, the detailed terminology and definitions, and the extensive appendices all add up to a Code that faces the same complexities resulting from past procedures that must be protected, and is thus quite as complicated as the Botanical Code, if not more so.

It is, however, true that the botanical tradition has been to amend the Code rather than authorize exceptions. The transatlantic opposition to the first list of *nomina generica conservanda* of the Vienna Code (see above) and the later discovery (Rickett and Stafleu 1959-1961) that very many names on the list did not, in fact, require conservation, have probably strengthened this point of view. For example, when it was realised at the Seattle Congress in 1969 that adoption of the first lectotypifications of several generic names that had been made by Britton and Brown (1913) under the American Code would lead to undesirable changes in the application of some of these names, the favoured approach (Stafleu and Voss 1972 pp. 25-28) was to amend the relevant Article (to proscribe lectotypifications said to be "made arbitrarily") rather than to deal with the problem directly by conserving the names with more appropriate types.

There is another facet to the tension between a simple code and a complex one. Some believe that nomenclature is best served by a degree of generality, even imprecision, in the Code, so that individual cases may be judged on their merits (i.e. with the desire to preserve current usage overriding consistency of application), without the rigidity of a detailed code that might force disturbance of existing usage. Others emphasise the nomenclatural instability that divergence of interpretation of a more general code may permit, and seek, for that reason, to make the Code cover almost all eventualities. Fortunately, Nomenclature Sections of Botanical Congresses have rarely been sympathetic to amendments designed to deal with situations of extreme rarity.

Likewise nomenclaturalists are divided between those who believe that the rules should always be strictly applied without exceptions and regardless of the name changes that this may induce, as against those to whom the maintenance of existing usage is of paramount importance. Although this is conceptually linked to a complex code, which covers most eventualities, versus a simple one, the truth is that this appears to be more a matter of human psychology than of the complexity of the code concerned. If anything, the U.S. draft code of 1905 and the American Code of 1907 (Arthur *et al.* 1905, 1907) were simpler documents than the Vienna Code (Briquet 1906), yet exceptions through conservation were anathema to the Americans but acceptable *en masse* to the botanists at Vienna. Incidentally, we believe this may be less a matter of a different North American psychology, as of the fact that the European flora has a much higher proportion of 18th century names, with their greater problems of typification, and hence the greater chance of latter-day upsets in well-established nomenclature. Given this legacy, Europeans are thus more ready to adopt special measures to minimize its disruptive effects.

As noted above, both the Botanical and Zoological Codes are relatively complex documents, yet there is no question but that the majority of botanical nomenclaturalists have been less willing than their zoological counterparts to legislate for machinery to allow individual names to be exempted from the normal workings of the codes. Although *nomina generica conservanda* date to the Vienna Code, *nomina familiarum conservanda* were first accepted only in Montréal in 1959; *nomina rejicienda*, without corresponding *nomina conservanda*, were provided for in Leningrad in 1975, whereas *nomina specifica conservanda* finally gained the necessary 60% acceptance by only 3 votes at the IBC Nomenclature Section meeting in Sydney in 1981   and that only for species of "major economic importance". It is probably safe to say that the memory of the American Code schism has been an important factor in the different approach of botanical nomenclaturalists to that of their zoological counterparts.

A rather different tension existing in biological nomenclature is between historical propriety and the achievement of stability by arbitrary decree. There are those who, in considering typification for example, lay great store by historical accuracy: what really was the dominant element in an eighteenth century author's mind when he prepared his, by modern standards, totally inadequate diagnosis of a new taxon?   Others take the view: why spend months studying obscure sixteenth or seventeenth century books, as well as the man's mode of work, his secret codes, and even his psychology, just to determine the application of one untypified name;   much better to make a quick, arbitrary decision in a manner that conforms to most current usage   and then get on with some real science!   This sort of conflict is implicit in all discussions of lectotypification and neotypification, including the extent to which priority of lectotypification is inviolate, and the circumstances under which neotypification should be permissable. Both points of view are well-represented in contemporary botanical nomenclature, and, for example, tended to be opposed in the debates of the Special Committee on Generic Typification which reported to the Sydney Congress (McNeill 1981). One group, to which the name "idealistic" was applied, believed that the material that the author actually studied must take precedence, whereas the other, the so-called "pragmatic" school, believed that where there was a discrepancy, the identity of the type of the name of the author's cited "type species", should determine the application of the generic name. In the event the "pragmatic" view prevailed (Greuter and Voss 1982 pp. 40-49)[5], but probably more because the proposals of that group were more convincingly worded, than because of innate preference by the Nomenclature Section.

Botanical nomenclature has its share of those who espouse each of these philosophical positions. A nomenclatural code that is to be acceptable to the great majority of taxonomists must reflect a balance of interests amongst the different viewpoints held by its users. The need for a clear majority vote of a broadly based structure such as the Nomenclature Section of a Congress helps to ensure that this compromise and consensus will be maintained. Although rapporteurs, such as Briquet and Lanjouw, have at times sought to use their influence to change the Code for the better, as they perceived it, there has also been the tradition that part of the mandate of the rapporteurs is not only to advise the Nomenclature Section on the likely practical implications of proposals to amend the Code, but also to use their influence to ensure that botanical nomenclature continues to reflect a broadly based consensus. We hope that we shall continue that tradition.

Current Issues in Botanical Nomenclature

Despite the preliminary publication in *Taxon* of proposals to amend the Code, the comments of the rapporteurs, and the subsequent debate at the Congress, issues emerge at nearly every Congress, which cannot be resolved in the five days normally available for meetings of the Nomenclature Section. These issues are generally referred to Special Committees for report to the Nomenclature Section of the next Congress.

Four such committees are currently established (Voss 1982): one deals with criteria for effective publication, one with those for valid publication ("availability" under the ICZN), one with lectotypification, and one with orthography.

----

[5] A classic case is that of *Pseudolarix*, described by Gordon (1858) on the basis of plants of the Chinese golden larch, which he referred to as *P. kaempferi* (Lambert) Gordon, his "type species". However, the type of *P. kaempferi* is a true *Larix* and, indeed, *L. kaempferi* (Lambert) Carrière is the correct name for the Japanese larch. Thus Gordon's "type species" includes two discordant elements: the material he described and the material represented by the name he used. From which should the type be chosen? The Botanical Code now explicitly designates the latter, reflecting the "pragmatic" view, with provision for conserving the name with a type drawn from one of the specimens used by the author, if this best serves nomenclatural stability. It is this conservation solution that is being proposed, and will probably be adopted, in the *Pseudolarix* case (cf. Hara & Yü 1983).

The requirements for publication to be "effective" under the Botanical Code (equivalent to "publication" in the Zoological Code), have two different elements within them. First of all "printed matter" must be used, and, secondly, this must be distributed to "the general public or at least to botanical institutions with libraries accessible to botanists generally" (ICBN Art. 29.1) (cf. Weresub and McNeill (1980), Brummitt (1980), and Nicolson (1980) for analyses of these requirements). Technological developments in printing and publishing have blurred the old distinction between "printed matter" and "type-scripts or other unpublished material" (ICBN Art. 29.1). For example, reduced xeroxing of typescript or laser printout cannot (or can hardly) be told apart from the offset processes used by many important plant systematic journals such as *Phytologia, Mycotaxon*, and *Mitteilungen (aus) der Botanischen Staatssammlung München*. Moreover, future trends in printing and publishing are likely to make it even more difficult to maintain such a distinction. The matter of distribution is also being questioned. Strictly, for publication to be effective, it is only necessary to have proof of receipt of a "printed" sheet of paper by two "botanical institutions with libraries accessible to botanists generally". Apparently the sheets of paper need never even be deposited in the libraries! Some guarantee of greater availability seems desirable. Another problem arises from theses produced by microfilm-xerography as, for example, by University Microfilms of Ann Arbor, Michigan. These are often not intended to be effectively published by their authors, but, provided two libraries have copies, it is hard to see how they can be viewed otherwise under the present wording of the Code.

The second matter currently under review by a Special Committee is that of the requirements for valid publication under the Botanical Code (equivalent to "availability" in the Zoological Code). Compilers of indices of names, such as *Index Kewensis*, have found the present rules ambiguous, even to the point of providing their own unofficial guidelines for determining validity (Royal Botanic Gardens, Kew 1974). Much of this is concerned with how precise "a full and direct reference" to "author and place of valid publication with page or plate reference and date" (ICBN Art. 33.2) must be. Other matters that are of concern in determining whether a name is validly published relate to the distinction of meaning between verbs such as "cite", "designate", and "indicate", when applied, for example, to types.

Valid publication of names of new taxa under the Botanical Code on or after 1st January 1958, requires indication of the nomenclatural type. For many names published prior to that date, lectotypification is often necessary. The Code specifies that the first lectotype to be designated establishes the application of the name. The Code is not clear, however, as to what the minimum requirements are for an action to be considered lectotypification. Possibilities range from regarding the exclusion of some of the elements included by the original author as restricting choice to those retained, to requiring an evaluation of all potential types, prior to the designation of a single element.

The final topic that is currently the subject of consideration by a Special Committee is orthography. There are two perennial questions: (i) whether an author is deliberately using a particular spelling or whether he is just making a mistake, and (ii) what is, in fact, the grammatically most correct form, particularly with regard to the "stem" of a name, where there may be no classical usage or even a change of usage from classical to mediaeval times. One regularly canvassed solution is to insist on always using the original spelling; others find it repugnant to use forms that clearly stem from grammatical or other errors. The matter gains importance nowadays with the extensive development of electronic storage and retrieval of data. Without special programming, a computer will inevitably treat two spelling variants as two different names.

Future Directions of Botanical Nomenclature

The issues currently being addressed by the Special Committees are evidently ones that will be considered further at the Nomenclature Section of the next (the XIVth) International Botanical Congress, to be held in Berlin (West), Germany, from 20th to 24th July 1987, prior to the main Congress sessions. It is likely that some proposals to clarify criteria for valid publication and for priority of lectotypification will be proposed and probably accepted. Matters of orthography may prove less tractable.

The question of effective publication, however, may turn out to stimulate the greatest debate and, perhaps, the most significant progress. If this happens, it is likely to be in a direction quite different from any canvassed at the previous Congress in Sydney in 1981 (Greuter and Voss 1982, pp. 66-72).

Instead, it is possible that there will be a proposal to establish a registry of new names of taxa governed by the Botanical Code. Although this is not altogether a new idea (cf. Thomas 1976), the technological developments that might permit such a scheme to be established, practically and economically, have not hitherto existed. The stimulus for the proposal, whose feasibility is currently being studied by an *ad hoc* committee established by the General Committee, came from discussions at the 1985 International Congress of Systematic and Evolutionary Biology (ICSEB III) and later from the deliberations and resolutions of the XXII General Assembly of IUBS at Budapest in September 1985 (IUBS 1986). The first stage, if found feasible, would simply be to establish a registry to which authors of new names would be invited to submit documentation. If the registry proved an effective means of communication of the documentation on new taxa, the Nomenclature Section of a later Congress might be asked to require listing in the registry as a prerequisite for the valid publication of new names. The enormous information retrieval advantages of a registry of all new names, could be expected to offset any sense of restriction on freedom of publication, particularly as attempts would be made to ensure a decentralized registry.

Although the matter of organisms, such as protists, that come under more than one code has been addressed specifically by other contributors to this symposium, a few general remarks on directions which the ICBN may take to deal with this problem should be included.

There are a number of possible solutions. The most extreme is to release definable groups from the Code (are "protists" a definable group?). This has already been done with the bacteria (see above), but the determination of what constitutes a bacterium is a matter of taxonomy not nomenclature. It appears, for example, that most taxonomists working on the Cyanophyta/Cyanobacteria prefer to regard them as algae, at least for the purposes of nomenclature. This, of course, is in part because the Approved Lists of Bacterial Names do not provide for the "Cyanobacteria". Botanists would certainly not want to include groups of organisms within their code, if the specialists in those groups preferred a different arrangement *and* if an acceptable alternative code was clearly established and in operation. This is the history of the separation of the bacteria from the jurisdiction of the Botanical Code: the first edition of the Bacteriological Code appeared in 1958 (Buchanan *et al.* 1958), and, moreover, was based on a Code adopted in 1947 (Buchanan *et al.* 1948), but the removal of bacteria from the Botanical Code did not take place until the XII International Botanical Congress at Leningrad in 1975, by which time it had become clear that the Bacteriological Code was established and accepted by bacteriologists.

Alternatively, special requirements of a definable group could be met within the Botanical Code. This is currently the case with the "algae" and "fungi", which may not be holophyletic groups, but which are, apparently, still definable. The latter has a particularly well-developed special component (ICBN Art. 59), to cover fungi that are only known, or are only widespread, in one of the reproductive stages (anamorph or teleomorph) of the whole life history of the organism (holomorph) (cf. contribution by P.M. Kirk in this symposium).

We suspect that the best solution, however, is the one that Corliss (1983, 1984, 1986) has described as "futuristic". This requires protist systematists to develop a much more widely accepted, well-documented, and hence more stable system of higher classification of these organisms. Once their taxonomic relationships become more evident, it should be possible to agree on which groups (*regna, phyla* or *divisiones*) will fall under which code.

Meanwhile, it is more likely that either Code will be applied according to the preferences of individual workers and a case by case approach will be adopted to deal with discrepancies, so that, if at all possible, the same name will be correct under the Botanical Code and the Zoological Code (as well as the Bacteriological Code, if applicable).

So far we have not addressed, seriously, the question of harmonizing the Botanical and Zoological Codes in particular, and, by implication also the Bacteriological Code. We believe that this is something that should be addressed seriously. Our outline of the history and organization of botanical nomenclature makes it clear that simple solutions are impossible and that it would be naive to imagine that any progress in this direction is possible except where there are evident benefits to be achieved. One approach is to work together towards the development of new common strategies. The development of registries of new names, referred to above, is one such strategy. Others may be more mundane, as, for example, agreement on word endings signifying taxonomic ranks, and ensuring that such changes in the respective codes as are not peculiarly botanical or zoological (e.g. effective publication) are made in a compatible manner. Further, and perhaps more extensive, meetings of biologists interested in botanical and zoological nomenclature, held at each ICSEB, could facilitate this process.

Summary

The history of botanical nomenclature and its organizational structure are important factors in determining the form of the present *International Code of Botanical Nomenclature.* The Code has developed in response to particular situations in botany and often represents a compromise between different approaches to legislation. These tensions still exist in botanical nomenclature and there is ongoing debate on how best to improve the Code. It must be noted that any change in the Code, however small, almost inevitably leads to name changes. Consequently, harmonization with other codes, such as the International Code of Zoological Nomenclature, involving changes to the Botanical Code, is only possible in cases in which the benefits to be achieved are evidently substantial. We hope that the future will permit harmonization between codes in terms of developing new strategies to meet the challenges of modern technology and advances in systematic biology.

Acknowledgements

We are greatly indebted to F.A. Stafleu (Utrecht), E.G. Voss (Ann Arbor), K. Pryer (Ottawa), P.G. Parkinson (Wellington), A. Kanis (Canberra), D.H. Nicolson (Washington), and D.L. Hawksworth (CAB IMI, Kew) for their valuable comments on an earlier draft.

REFERENCES

Anonymous. 1892. Proceedings of the botanical club of the A.A.A.S. *Bot. Gaz.* 17: 285-290.

Arthur, J.C. *et al.* (Nomenclature Commision). 1904. Code of Botanical Nomenclature / Code de la Nomenclature botanique / Kodex der botanischen Nomenklatur. *Bull. Torrey Bot. Club* 31: 249-290.

Arthur, J.C. *et al.* (Nomenclature Commision). 1907. American Code of Botanical Nomenclature. *Bull. Torrey Bot. Club* 34: 167-178.

Bold, H.C., Cronquist, A., Jeffrey, C., Johnson, L.A.S., Margulis, L., Merxmüller, H., Raven, P.H., and Takhtajan, A.L. 1978. Proposal (10) to substitute the term "Phylum" for "Division" for groups treated as plants. *Taxon* 27: 121-122.

Brickell, C.D., Richens, R.H., Kelly, A.F., Schneider, F., and Voss, E.G. (eds). 1980. *International Code of Nomenclature of Cultivated Plants 1980.* Bohn, Scheltema and Holkema, Deventer, Netherlands. 32 pp. (*Regnum Veg.* 104).

Briquet, J. 1906. *Règles internationales de la Nomenclature Botanique adoptées par le Congrès International de Botanique de Vienne 1905.* Gustav Fischer, Jena. 99 pp.

Briquet, J. 1935. *International Rules of Botanical Nomenclature . . . . Revised by the International Botanical Congress of Cambridge, 1930.* Gustav Fischer, Jena. xi + 151 pp.

Britton, N.L., and Brown, A. 1913. *An illustrated flora of the northern United States, Canada and the British possessions.* Ed. 2. 3 vols. Scribner's, New York.

Brummitt, R.K. 1980. Questions of effective publication. *Taxon* 29: 477-483.

Brummitt, R.K. 1981. Report of the Special Committee on Autonyms. *Taxon* 30: 183-200.

Buchanan, R.E., Cowan, S.T., Wikén, T. and Clark, W.A. (eds.). 1958. *International Code of Nomenclature of Bacteria and Viruses.* Iowa State College Press, Ames, Iowa. 186 pp.

Buchanan, R.E., St. John-Brooks, R. and Breed, R.S. 1948. International bacteriological code of nomenclature. *J. Bact.* 55: 287-306.

Bureau of Nomenclature [Stafleu, F.A.]. 1960. IX International Botanical Congress Nomenclature Section. Report. *Regnum Veg.* 20. (Reprinted with separate pagination from *Proceedings of the IX International Botanical Congress, Montréal 1959,* 3: 27-116. Univ. of Toronto Press).

Candolle, Alph. de 1867. *Lois de la Nomenclature Botanique, adoptées par le Congrès International de Botanique tenu à Paris en août, 1867, suivies a une deuxième édition de l'introduction historique et du commentaire qui accompagnaient la rédaction préparatoire présentée au Congrès.* H. Georg, Genève and Bale; J.-B. Baillière et fils. Paris. 64 pp. (The *Lois* also published as pp. 209-255 in Fournier, E. 1867. *Actes du Congrès International de Botanique tenu à Paris en août 1867.* G. Baillière, Paris).

Candolle, Alph. de 1883. *Nouvelles remarques sur la nomenclature botanique.* H. Georg, Genève. 79 pp.

Cheeseman, T.F. 1908. Notes on botanical nomenclature, with remarks on the rules adopted by the International Botanical Congress of Vienna. *Trans. Proc. New Zealand Inst.* 40: 447-465.

Corliss, J. 1983. Consequences of creating new kingdoms of organisms. *BioScience* 33: 314-318.

Corliss, J. 1984. The kingdom Protista and its 45 phyla. *BioSystems* 17: 87-126.

Corliss, J. 1986. Progress in protistology during the first decade following the reemergence of the field as a respectable interdisciplinary area in modern biological research. *Progr. Protistol.* 1: (in press).

Cowan, R.S. and Stafleu, F.A. 1982. The origins and early history of I.A.P.T. *Taxon* 31: 415-420.

Fairchild, D.G. 1892. Proceedings of the botanical club of the forty-first meeting of the A.A.A.S., Rochester, N.Y., August 18-24, 1892. *Bull. Torrey Bot. Club* 19: 281-298.

Fournier, E. 1867. Discussion des Lois de la Nomenclature botanique. Pages 177-205 in: *Actes du Congrès International de Botanique tenu à Paris en août 1867.* G. Baillière, Paris.

Gordon, G. 1858. *The Pinetum*. H.G. Bohn, London. 353 pp.

Greuter, W. and McNeill, J. 1981. Proposal to permit conservation of species names. *Taxon* 30: 288.

Greuter, W. and McNeill, J. 1983. Concerning amendment proposals. *Taxon* 32: 660-661.

Greuter, W. and Voss, E.G. 1982. Report on botanical nomenclature Sydney 1981. *Englera* 2: 1-124.

Hara, H. and Yü, T.T. 1983. (709) Proposal to conserve 25 *Pseudolarix* Gordon (Pinaceae) with a type specimen. *Taxon* 32: 485-487.

Hitchcock, A.S. 1919. Report of the Committee on Generic Types of the Botanical Society of America. *Science* 49: 333-336.

Hitchcock, A.S. 1921. Report of the Committee on Nomenclature of the Botanical Society of America. *Science* 53: 312-314.

Hitchcock, A.S. 1925. *Methods of descriptive systematic botany*. John Wiley, New York; Chapman & Hall, London.

IUBS 1983. International Union of Biological Sciences / Union internationale des Sciences biologiques. XXIst General 'Assembly. Ottawa, Canada, 22-27 August, 1982. *IUBS Publ Sér. A (Générale)*, 21. (ISBN 92 9046 0695).

IUBS 1986. International Union of Biological Sciences / Union internationale des Sciences biologiques. XXIInd General Assembly. Budapest, Hungary, 1-8 September, 1985. *IUBS Publ Sér. A (Générale)*, 22.

Jeffrey, C. 1977. *Biological Nomenclature*. ed. 2. Edward Arnold, London. viiii + 72 pp.

Kuntze, O. 1891-1898. *Revisio generum plantarum*. A. Felix, Leipzig; Dulau and Co., London; U. Hoepli, Milano; G.E. Schechert, New York; C. Klincksieck, Paris; G. Robertson and Co., Melbourne. 3 vols.

Lanjouw, J. 1959. *Synopsis of proposals concerning the International Code of Botanical Nomenclature submitted to the Ninth International Botanical Congress, Montréal 1959*. International Association for Plant Taxonomy, Utrecht. 84 pp. (*Regnum Veg.* 14).

Lanjouw, J. (ed.). 1950. Botanical Nomenclature and Taxonomy. (IUBS Publ. Ser. B Colloquia, 2). *Chronica Botanica* 12: 1-87.

Lanjouw, J., Baehni, Ch., Merrill, E.D., Rickett, H.W., Robyns, W., Sprague, T.A. and Stafleu, F.A. (eds). 1952. *International Code of Botanical Nomenclature. Adopted by the Seventh International Botanical Congress, Stockholm, July, 1950*. Chronica Botanica. Waltham, Mass., U.S.A. 228 pp. (*Regnum Veg.* 3).

Lanjouw, J., Baehni, Ch., Robyns, W., Ross, R., Rousseau, J., Schopf, J.M., Schulze, G.M., Smith, A.C., Vilmorin, R. de, and Stafleu, F.A. 1961. *International Code of Botanical Nomenclature. Adopted ... Montréal, August 1959*. International Association for Plant Taxonomy, Utrecht. 372pp. (*Regnum Veg.* 23).

Lanjouw, J., Mamay, S.H., McVaugh, R., Robyns, W., Rollins, R.C., Ross, R., Rousseau, J., Schulze, G.M., de Vilmorin, R. and Stafleu, F.A. 1966. *International Code of Botanical Nomenclature. Adopted by the Tenth International Botanical Congress, Edinburgh, August, 1966*. International Association for Plant Taxonomy, Utrecht. 402 pp. (*Regnum Veg.* 46).

Lapage, S.P., Sneath, P.H.A., Lessel, E.F., Skerman, V.B.D., Seeliger, H.P.R., and Clark, W.A. (eds.). 1975. *International Code of Nomenclature of Bacteria.* American Society for Microbiology, Washington D.C. 180 pp.

Lawrence, G.H.M. 1951. *Taxonomy of vascular plants.* Macmillan, New York. xiii + 823 pp.

Linnaeus, C. 1737. *Genera plantarum.* Conrad Wishoff, Lugduni batavorum [Leiden].

Linnaeus, C. 1753. *Species plantarum.* Laurentius Salvius, Holmiae [Stockholm]. 2 vols. 1200 pp.

Little, E.L. jr. 1957. Three proposals toward stabilization of botanical nomenclature. *Taxon* 6: 189-194.

McNeill, J. 1981. Report of the Committee on Generic Typification. *Taxon* 30: 200-207.

McVaugh, R., Ross, R., and Stafleu, F.A. 1968. *An annotated glossary of botanical nomenclature.* International Association for Plant Taxonomy, Utrecht. 31 pp. (*Regnum veg.* 56).

Melville, R.V. 1986. Some aspects of zoological nomenclature. *Biology International* Special Issue 9:

Nicolson, D.H. 1980. Key to identification of effectively / ineffectively published material. *Taxon* 29: 485-488.

Parkinson, P.G. 1975. The International Code of Botanical Nomenclature: an historical review and bibliography. *TANE: J. Auckl. Univ. Field Club* 21: 153-173.

Rendle, A.B. 1905. The botanical congress at Vienna. *Nature* 72: 272-274.

Rickett, H.W. and Stafleu, F.A. 1959-1961. Nomina generica conservanda et rejicienda spermatophytorum. *Taxon* 8: 213-243, 256-274, 282-314; 9: 67-86, 111-124, 153-161; 10: 70-91, 111-121, 132-149, 170-193.

Ride W.D.L., Sabrosky, C.W., Bernardi, G., and Melville, R.V. (eds.). 1985. *International Code of Zoological Nomenclature.* Ed. 3. University of California Press, Berkeley, Calif. 338 pp.

Ride, W.D.L. 1984. On the organisational and financial arrangements of organisations responsible for botanical and zoological nomenclature. *Taxon* 33: 240-246.

Ride, W.D.L. 1986. Zoological nomenclature. *Biology International* Special Issue 9:

Royal Botanic Gardens, Kew. 1974. The Kew Record, a new annual bibliography. *Taxon* 23: 381-386.

Smith, A.C. 1957. Fifty years of botanical nomenclature. *Britonnia* 9: 2-8.

Stafleu, F.A. and Voss, E.G. 1972. *Report on botanical nomenclature Seattle 1969.* A. Oosthoek, Utrecht. 133 pp. (*Regnum Veg.* 81)

Stafleu, F.A. 1985. Joseph Lanjouw (1902-1984). *Taxon* 34: 1-4.

Steenis, C.G.G.J. van 1957. Specific and infraspecific delimitation. In C.G.G.J. van Steenis (ed.), *Flora Malesiana.* Ser. 1, 5: clxvii-ccxxxiv. Nordhoff, [Djakarta/Leyden].

Thomas, J.H. 1974. New names etc. to be published in a single journal. *Taxon* 23: 833-835.

Voss, E.G. 1969. Nomenclature. Pages 730-731 in *Encyclopaedia Britannica* vol. 21 (subheading under "Taxonomy").

Voss, E.G. 1982. Announcement: Nomenclature Committees. *Taxon* 31: 718.

Voss, E.G. and Greuter, W. 1981. Synopsis of proposals on botanical nomenclature, Sydney 1981. *Taxon* 30: 95-141.

Voss, E.G. 1979. Twelfth International Botanical Congress, Leningrad, 1975. Nomenclatural Section Report. Pages 129-186 in *Proceedings XII International Botanical Congress, Leningrad 1975.* Science [Nauka], Leningrad.

Voss, E.G., Burdet, H.M., Chaloner, W.G., Demoulin, V., Hiepko, P., McNeill, J., Meikle, R.D., Nicolson, D.H., Rollins, R.C., Silva, P.C. and Greuter, W. (eds). 1983. *International Code of Botanical Nomenclature. Adopted by the Thirteenth International Botanical Congress, Sydney, August, 1981.* Bonn, Scheltema and Holkema, Utrecht/Antwerpen; W. Junk, The Hague/Boston. xv + 472 pp. (*Regnum Veg.* 111).

Weresub, L.K. and McNeill, J. 1980. Effective publication under the Code of Botanical Nomenclature. *Taxon* 29: 471-476.

Weresub, L.K. 1970. Automatic tautonyms: Zoological vs. Botanical Code. *Taxon* 19: 787-788.

Appendix: Abbreviations used in text

AAAS	American Association for the Advancement of Science
IABMS	International Association of Botanical and Mycological Societies (see page for membership)
IAPT	International Association for Plant Taxonomy
IBC	International Botanical Congress
ICBN	International Code of Botanical Nomenclature
ICNB	International Code of Nomenclature of Bacteria
ICSEB	International Congress of Systematic and Evolutionary Biology
ICNCP	International Code of Nomenclature of Cultivated Plants
ICZN	International Code of Zoological Nomenclature
IUBS	International Union of Biological Sciences

ZOOLOGICAL NOMENCATURE

W.D.L. Ride

When introducing the English version of the Règles Internationales de la Nomenclature Zoologique (adopted by the 6th International Congress of Zoology at Berne in 1904), Dr. Charles Wardell Stiles, one of the Commissioners appointed by the International Congress of Zoology to develop it said:

> The known genera and species of animals represent but a fraction (but ten to twenty percent) of the zoological names which will come into use during the next two or three centuries. It is clear that our nomenclatural tasks are easy, compared with the tremendous number of technical names the future generations will fall heir to. Under these circumstances, it is seen that in order to prevent our science from becoming "a mere chaos of words," every zoological author owes a serious nomenclatural duty, not only to himself and his colleagues of today but also to future generations of zoologists. If it were left to each author to accept or reject names according to his own personal wishes in the matter, the science of zoology would soon reach a stage in which it would be difficult for one author to understand the writings of another, hence in order to prevent such a chaotic state, systematists have felt themselves forced to adopt certain rigid rules in accordance with which any given animal has only one valid name, and that name shall be valid not only in the country in which it is proposed, but in all other lands as well." (Quoted by D.S. Jordan, 1905)

The task facing zoologists now is not very difficult. Today it is estimated that there are probably more than 1 000 000 species of living animals[2] (Rothschild, 1965). Mayr (1969), in describing the task of the taxonomist, notes that estimates of the total number of species (known and unknown) are as high as between 5 and 10 million (and of 50 to 100 times as many fossil species). In 1953, Mayr, Linsley and Usinger estimated that the rate of description of species was about 10 000 per year. The task ahead is immense. For this purpose the Code, as a working tool, must be efficient. It must be relevant. It must be supported and followed universally.

----

[1] Dr. Ride is President of the International Commission on Zoological Nomenclature. He was Chairman of the Editorial Committee of the 3rd Edition of the International Zoological Nomenclature.

[2] Compared with currently more than 371 500 species of plants (Wagenitz, 1967).

Aim and Key Elements of the Zoological Code

The stated aim of the International Code of Zoological Nomenclature (3rd Edn.) is to provide for maximum universality and continuity in use of scientific names for animals compatible with the freedom of scientists to classify animals according to taxonomic judgements.

The Code consists of mandatory rules, and recommendations for guidance. Together these are designed to enable zoologists to name any animal correctly either from pre-existing names, or, if necessary, by giving it a new name. By means of the Code, the valid name can be given at any rank from subspecies to superfamily.

The key elements of the modern Code are enumerated in the Introduction to the 3rd Edition (Ride, 1985). They are:

--- The Code does not determine the rank of any group of animals but, rather, provides the name to be used for the taxon at whatever rank it is assigned.

--- Names are allocated to taxa without interfering with the freedom of scientists to classify animals according to taxonomic judgements. This is achieved by means of the "device of types" (called "name-bearing types" in the Code to avoid confusion with other kinds of "types"). Every name covered by the Code is permanently attached to a name-bearing type. At the ranks of species and subspecies this name-bearing type is a single specimen (or a group of specimens that collectively constitute the name-bearer). At the ranks of genus and subgenus it is a nominal species (see Melville, Jeffrey, this volume); at ranks between tribe and superfamily it is a nominal genus. Accordingly, if a taxon, as assembled by taxonomy, contains several name-bearing types (each with a name that is available for use at that rank), the Principle of Priority operates to determine which of the synonyms (if there is more than one name) is the valid name of the taxon.

--- The Code recognizes that the Principle of Priority may occasionally upset a long-accepted name as the result of the exhumation of a little known, or even long-forgotten, name. In such cases the International Commission on Zoological Nomenclature is given plenary power to set aside the automatic operation of the Code to preserve the name that is threatened.

--- To avoid ambiguity, the Code prohibits the use of the same name (homonyms) for different taxa.

--- The Code provides guidance for zoologists needing to establish new names. The Code provides rules to determine whether any name, previously proposed, is available for use in zoological nomenclature, and with what priority; whether it required amendment for its correct use, and for its name-bearing type to be identified (and, when none can be discovered, to establish one).

--- The Code also provides for it to be interpreted and administered by the International Commission on Zoological Nomenclature, and also for it to be up-dated and improved.

Finally, there is no "case law" in zoological nomenclature. In most cases, questions may be settled from the Code alone - never by precedent. If the Commission is called on by a zoologist to give an Opinion or issue a Direction in a case not covered automatically by the Code, the decision of the Commission relates to that case alone.

History of Zoological Nomenclature

The origin of an internationally accepted Code of Zoological Nomenclature may be found in the confusion of names that occurs in the zoological literature of the early part of the 19th century. Following the publication of the 10th edition of the "Systema Naturae" by Linnaeus in 1758, and his adoption in it of consistently binominal names for species of animals, the next century saw the new system expanded and developed in different places, and in different ways for different animal groups. By the second quarter of the 19th century, disparate usages were common and the need for regulation was felt widely. Moreover, the great explosion in known species, resulting from the growth of science and from active scientific exploration in countries outside Europe, resulted in a multiplicity of names and of synonyms.

The most effective of the early attempts to regulate zoological nomenclature was developed by Hugh Edwin Strickland. A committee including such distinguished zoologists as Strickland, Charles Darwin, Richard Owen, and J.O. Westwood, made a report to the British Association for the Advancement of Science in 1842 entitled "Series of Propositions for Rendering the Nomenclature of Zoology Uniform and Permanent". That report was published (Strickland et al., 1843), translated and circulated widely. It had great influence. It was published in France, in Italy, and in the United States of America. It was adopted by the Scientific Congress at Padua in 1843, by the American Society of Geologists and Naturalists in 1845, and by the British Association for the Advancement of Science in 1846. It was revised in succeeding years and provided the basis for the Douvillé Code (1882) adopted internationally by geologists, and of the American Ornithologists' Union Code (1886).

Following discussion at International Congresses of Geology (Paris, 1878, and Bologna, 1881) it became clear that international agreement for rules to cover all zoological names was desirable. At the First International Congress of Zoology (Paris, 1889), the Congress adopted, in part, rules drawn up by Professor Raphael Blanchard and referred the matter for discussion at the Moscow Congress (1892). The Leyden Congress (1895) appointed an International Commission to report to the succeeding Congress (Cambridge, England, 1898). Following further consideration, publication was authorized by the Berne Congress (1904). The resultant code was published in French, English and German in 1905. This code entitled "Règles Internationales de la Nomenclature Zoologique", with a series of amendments resulting from subsequent Congresses (Boston, 1907; Monaco, 1913; Budapest, 1929; Padua, 1930) remained in force until 1961 when it was replaced in its entirety by the first edition of the "International Code of Zoological Nomenclature" that was the product of the Congresses following the 1939-45 War (Paris, 1948; Copenhagen, 1953; London, 1958 - see Stoll, 1961).

At the last held International Congress of Zoology XVII, Monaco, 1972) it was decided to transfer responsibility for future Codes (and the Commission) to the International Union of Biological Sciences (IUBS) and, thereby, to ensure mechanisms for continuity and future up-dating in the even that there were no further International Congresses of Zoology. The transfer was accepted by IUBS at the XVIII General Assembly of IUBS (Ustaoset, Norway, 1973).

The modern "International Code of Zoological Nomenclature" (3rd Edn.) (ICZN, 1985) is complex, partly because of the interdependence of different parts of it, and partly because it must contain devices to validate actions and names in use but resulting from earlier, less satisfactory, standards of description and publication. Many of these names are products of the unregulated period before the Règles. For instance, the principle of name-bearing types, seen to be so fundamental today to the objective identification of names and for determining the correct names for taxa, was not made explicit in the rules for names of genera and subgenera until the Budapest Congress in 1927 (although it is implicit in Articles governing the formation of names of families and subfamilies (Art. 4) and for the division of genera (Arts. 29, 30) that type-genera and type-species were in operation from the Règles - see Stiles, 1905). Even today, although the principle of types is fundamental, it is still not obligatory for types to be designated when new names are proposed for species and subspecies. However, since the Code required every new species-group name to be based upon a description of an actual specimen (or specimens), each name has, ipso facto, a type. Types are obligatory for genera, and the Code recommends the practice of explicitly designating types for species and subspecies. Where no type is nominated, the Code provides procedures by which the name-bearing type of the name of any species or subspecies can be discovered or fixed.

Updating and Improving the Code

No code of biological nomenclature is perfect. Moreover, the requirements of science change. Accordingly, the International Code of Zoological Nomenclature contains provisions for its amendment and for the participation of biologists in the amending processes, as well as in the ratification of amendments.

It is important to emphasize that whatever is done to amend the Code must be done with the fact in mind that no person or jurisdiction is bound by it; its provisions are mandatory only to the extent that zoologists are willing to observe them. They must be voluntarily acceptable. There can be no penalties for those who disregard them except those of conscience and the disapproval of colleagues. The Code cannot be amended in advance of general opinion among its users. Accordingly, the main force of the mechanisms for amending the International Code of Zoological Nomenclature is to prevent hastily conceived or insufficiently considered changes (See Ride, 1984, for a comparison between botanical and zoological mechanisms).

In addition, like an organism, the Code is an evolving entity that must remain viable at all stages of its evolution. Changes cannot be pushed too far or too fast. Moreover, the aim to provide stability and universality in nomenclature is not promoted by an unstable code. Finally, no change can ever allow nomenclature to infringe upon the taxonomic judgement of the user, and should not upset the stability and universality of currently accepted names.

To ensure participation by zoologists, the International Code of Zoological Nomenclature may only be amended by IUBS acting on a recommendation from the International Commission on Zoological Nomenclature of IUBS. Henceforth, meetings of the Section will take place not only at General Assemblies of IUBS but also, with the concurrence of the Commission, at international congresses convened by meetings of Scientific Members of IUBS such as ICSEB - see article this volume, and also Statutes of the Section of Zoological Nomenclature, Bulletin of Zoological Nomenclature (42:321-323).

Any proposal to amend the Code must be published at least 12 months before a meeting of the Section at which it is to be considered so that zoologists everywhere have an opportunity to comment in writing on the proposal before a vote of the Commission is taken. If the Commission makes a recommendation, it must then be considered by the Section and a recommendation made by the Section to IUBS. Such an arrangement allows every National and Scientific Member of IUBS to brief delegates to General Assemblies and to open discussion on at least two seperate occasions. Moreover, it ensures that before any decision is taken, there is very careful consideration by the Commission itself (to ensure that integration of the proposed amendment within the Code will not present problems) before zoologists become bound by it (ICZN 1985, p. 245). In future, special committees with responsibility in different parts of the animal kingdom will also review proposals, as well as advising the Commission on cases involving the plenary power (see Ride, this volume).

Pressures for Change

Science, and the social and technical systems within which scientists, work are progressively changing. These changes bring pressures onto the Code to accomodate new taxonomic practices and even such matters as administrative and legislative needs independent of taxonomy. In addition, there is always pressure from those who use the Code to make it more workable; to simplify it, to integrate its provisions, and to rearrange, and generally fine-tune its wording and structure.

Operational Improvements

Currently pressures for change are being applied from several directions. Firstly, improvers of its operation are seeking automatic solutions to issues such as those resulting from the discovery that stability of a name will be upset because its type (species or specimen) has been misinterpreted in the past. Currently, such issues are only solved by application to the Commission for it to intervene and, when necessary, to use its plenary power.

A second area of pressure from operational improvers lies in the scholastic adherence in the Code to the extent that names must conform to the requirements of Latin grammar of which few zoologists today, or in the future, can be expected to have any understanding. Most zoologists find these requirements unnecessarily burdensome. With increasing use of automatic data retrieval systems that do not cope easily with mandatory changes in termination of specific names to conform with different genders of generic names, and the abandonment of classical languages, the issue of amending the Code to terminate these requirements will not be long put off.

Finally, an increasing number of users of names of organisms in more than one kingdom (such as ecologists working with entire biotas, protistan workers with organisms treated under both botanical and zoological codes (see Ride, 1982; Corliss, 1984; Kirk & Hawksworth, this volume), and educators seek harmonization between the several codes. Petty differences between the products of the different codes (such as in the treatment of the form of names, citation of authorship, and the different use of parens) are a nuisance to scientists and editors alike. Moreover, with an increased frequency of multidisciplinary academic courses the different terminologies used by the botanical and zoological codes (such as of epithet, availability, validly published, legitimate, valid, specific name, name of species) are an unnecessary barrier to comprehension by students. Currently there is an awareness among those responsible for the codes of the need to remove such differences, but no formal proposals have yet emerged other than attempts that failed to introduce the term epithet into the zoological code and to dispense with "nominal taxon".

Unfortunately, there is no possibility of a single code to serve both botany and zoology (see Jeffrey, this volume). Studies made in recent years have revealed that fundamental differences, especially those of determining precedence that result from the principle of coordinate names in the zoological and bacteriological codes and its absence from the botanical code, would, if abandoned by one or the other, demand changes in a very high proportion of specific and subspecific names (i.e., at the levels most requiring stability). Fortunately, the principal impact of those differences in the codes is upon the people who are best able to cope with their inconvenience (i.e., nomenclaturists), and there would be a far greater inconvenience to science, generally, from the wholesale name-changing that would accompany unification. The International Union of Biological Sciences has supported an ecumenical approach to these issues (IUBS, 1982, Resolution 5 - "Common Approaches to Biological Nomenclature").

The most difficult of the operational issues that faces zoological nomenclature today is whether publication as the primary determinant of the availability of names is still appropriate. As could be expected, the Règles required publication (Art. 25, "was published and accompanied...") because at that time most scientific information was distributed and recorded in published works that were typeset. But technology changed and in 1948 the Paris Congress found it necessary to restrict publication to reproduction in ink on paper, a requirement that was incorporated in the 1961 edition of the Code. In the third edition (1985), the Commission, now faced with new methods of publishing and printing, unknown twenty years before, found the requirement for ink on paper to be unduly restrictive and it was removed from 1986. However, the removal, while admitting such techniques as microfiche and high quality printing by electrostatic processes, also opens the door to "photocopies" that, potentially, would make the operation of the principle of priority even more difficult than it has been in the past.

It remains to be seen whether the solution of admitting "unconventional printing", with safeguards proposed in the current edition, will prove satisfactory and whether zoologists, on the wole, will ignore such means of "easy" publication as beneath dignity (as they mostly disregarded hectographing and mimeographing, methods permissible under the earlier "ink on paper" requirement). But it seems likely, in the longer term and with the development of new information systems, that the solution will not lie in patching up a definition of publication but, rather, in scrapping it and finding a new means of dealing with availability.

Already it has been proposed to IUBS by the 3rd International Congress of Systematic and Evolutionary Biology (see article this volume) that a solution may be found in some process of registering new names (say, in the Zoological Record), or of confining the publication of new names and acts to designated publications (such as scientific journals), designated irrespective of the methods used in their production (see IUBS General Assembly Resolution 3, item 4, this volume).

Improvements Sought in Support of Changing
Taxonomic Procedures and Theory

Because of its simplicity and its wide acceptance over more than two centuries, the binominal system is robust and has survived several major revolutions in taxonomic theory. The most serious of these was the even of evolutionary biology.

Although the Linnaean binominal reflected essentialist philosophy (as well as providing a system of brief, convenient and informative names for application to individual organisms), it survived the shift to an empirical taxonomy based upon evolutionary principles, a taxonomy of variable populations with conceptually perceived boundaries. It survived without any real threat to it.

During the shift, the practical need to determine unambiguously, the allocation of names when formerly wider genera or species were split, led to the device of types (e.g., Strickland et al., 1843). By this means, when a genus is split, its name remains in use and goes to that part of the former genus that contains its type. It is a consequence of this indissoluble connection between a name and its type that boundaries of taxa may be expanded or contracted to express different taxonomic concepts independently of nomenclature. Such a system fulfilled the needs of the new empirical taxonomy very well. Subsequently, the addition of trinominals gave greater flexibility to deal with geographical variation and similar evolutionary situations (see Coues et al., 1892, pp. 2, 29-32).

With increasing interest in taxonomic lineages, proposals have been made from time to time to replace the binominal system by accretionary systems of naming. These are mostly numerical but some have been uninominal. They are proposed by authors regarding them as more informative about lineages and in particular about the location of taxa within lineages. However, such proposals have not been a serious challenge. Apart from the fact that the replacement of existing names would be a huge task, and numerical identifiers are non-memorable, such systems suffer from the disadvantage that a direct relationship between the name and the highly controversial information that it is designed to convey would produce names that are unstable by comparison with the names of the Linnaean hierarchy allocated through the type principle and independent of taxonomy. Wiley (1979) has reviewed such proposals.

The present revolution in taxonomic methodology produced by the extensive development of cladistic taxonomy, might be expected to produce pressures for change. In my view the only effect on nomenclature is a consequence of the tendency to reduce the size of genera as a result of restricting generic boundaries to monophyletic groups (in the sense of Hennig, i.e., to the exclusion of paraphyletic taxa unless all members remain included). The consequences of reducing the size of genera are to reduce the informative value of the binominal and to decrease stability of combinations.

The categories subgenera, species-groups, and superspecies, which may now be expressed in nomenclature by interpolating additional names if desired (see 3rd Edn. of the Code), are available for use in conveying such information while retaining the nomenclatural stability of large genera. Concurrently with cladistic pressures to reduce the size of genera, the concept of the "genetic clock" is being used to argue that genus formation in lineages should be fixed at a level 10-20 million years before present, a level that would ensure that except in the case of very modern radiations most genera would have had little time to evolve many species.

De Smet (1974, An Introduction to New Biological Nomenclature) has proposed that the first named rank above species should be the family - but I do not know anybody who has taken that proposal seriously. (I suspect that the fact that the proposal is embedded in a more general proposal to replace Latin with Esperanto may have caused it to become overlooked.) However, if in future it should be considered that the utilitarian value of the binominal should be maintained, and that it is sensible to fix genus formation temporally, it may be that taxonomic acceptance will grow for a concept that the hierarchical level currently assigned in taxonomy to the family could be that named by the first name of the binomen.

Improvements Sought by Non-Taxonomist Users of Names

Taxonomists are expected to exercise a responsibility to the ordinary users of binominals to retain stability by exercising restraint in proposing new classifications. It is not meant by this that they should falsify scientific information by erecting or perpetraing artificial groups in order to preserve combination, but that, in the past, except for workers on a few popular groups such as birds, taxonomists were mostly able to propose rival taxonomic hypotheses, and rearrange names to express these with thought only of the taxonomic debate that would be created. In recent years, ecology and conservation have produced a multitude of new users who are not taxonomists -and, in fact, many users are legislators far removed from science and who are intolerant of obscure scientific arguments that create instability in names that they have to rely on.

In the recent past (ICZN, 1972) we have seen how ornithologists in particular, responding to popular pressure, led the arguments that resulted in Articles 23b and 79c (and its forerunnner the so-called "statute of limitation" - also Article 23b - of the period 1961-1973), and in the provisions to restore erstwhile secondary homonyms. Today the requirements of international agreements including as the CITES international protocol, and bilateral agreements providing protection of migratory organisms, and legislation covering named species for environmental conservation, fisheries, and whaling are beginning to produce even more stringent pressures for increased stability.

It is important to realize that these pressures are equally pressures to find solutions that will allow taxonomy to function indepentently of nomenclature, as they are pressures to facilitate preservation of established usage threatened by nomenclatural upsets. Even in well-known groups today, by far the greatest number of upsets in names are caused by taxonomic changes (i.e., revised hypotheses of relationships) than are caused by the discovery of earlier names. In the last but one volume of Peters' Checklist of the Birds of the World more than 100 names were changed for taxonomic reasons (pers. comm., G.F. Mees).

Another factor that must be noticed is that because technical advances in publishing have reduced the cost and increased the ease and speed whereby new names may be established, stability in nomenclature has become very vulnerable to deliberately upsetting actions. Today irresponsible work can be published on a scale never before experiences (Tyler, 1985).

In the introduction to this paper, I mentioned that it was a fundamental principle of the Code that none of its provisions should impose any restraint upon taxonomy. Freedom to practice irresponsible and poor taxonomy has always been regarded as a matter to be dealt with by peer pressure. However, threats to stability at this new scale by actions taken validly under the Code (Thulborn, 1986) have resulted in a movement to seek to introduce into the Code some provision that would allow authors freedom to ignore works that are productive of confusion. A little reflection will show that a code that attempts to legislate for rejection of works or names by individual zoologists on the basis of taxonomic opinion would not work. Some would do one thing and others something else with the result that universality would be lost.

The proposal from ICSEB that registration of new names might be introduced could provide a solution to the dilemma, but a single registration authority for all names in zoology would be impossible to manage. However, a workable solution could be provided by requiring new names or acts affecting nomenclature to be published in specified refereed journals. Some would certainly object, regarding the process as censorship, but the refereeing process is widely accepted in science and, providing there is a wide enough choice of journals, such a requirement should not be over-restrictive within bounds of legitimate scientific speculation. Publication outside the authorized journals could be accepted or ignored according to scientific merit, but names in them would be "informal" until properly published in an approved work. The system currently in operation in bacteriology is of this sort, but because of the much smaller scale of bacteriology, operates through a single journal (see Sneath, this volume).

Acknowledgements

In developing the ideas expressed in this paper, I gratefully acknowledge the help of my colleagues on the former Editorial Committee of the 3rd Edition of the International Code of Zoological Nomenclature and, in particular, of L.B. Holthuis, R.V. Melville, and C.W. Sabrosky. J.O. Corliss stimulated a closer look at the problems of taxonomists dealing with Protists, and Ernst Mayr challenged and contributed to finding a solution to the problem of dealing with forgotten names. A.J. Cain and Francis Hemming started it all.

REFERENCES

CORLISS, J.O. 1984. The Kingdom Protista and its 45 phyla. Biosystems. 17:87-126

COUES, E. et al. 1892. The Code of Nomenclature Adopted by the American Ornithologists' Union. (2nd Edn) American Ornithologists' Union: New York. v, 72pp.

DE SMET, W.M.A. 1974. An Introduction to New Biological Nomenclature (NBN). Association for the Introduction of New Biological Nomenclature: Kalmthout (Belgium). 77pp.

ICZN (International Commission on Zoological Nomenclature). 1972. Minutes of the Meeting of the International Commission, 25 September, 1972...26 September, 1972...Section on Nomenclature, 28 Sept., 1972. Bull. Zool. Nomencl. 29:168-189.

ICZN. 1985. International Code of Zoological Nomenclature. Third Edition adopted by the XXth General Assembly of the International Union of Biological Sciences. International Trust for Zoological Nomenclature in Association with British Museum (Natural History), London. xx, 338pp.

IUBS. 1982. The XXIst IUBS General Assembly, 22-27 Aug., 1982, Ottawa, Canada. Biology International. N° 6:11-21.

JORDAN, D.S. 1905. (Review of) The International Code of Zoological Nomenclature as Applied to Medicine. Science. (n.s.) 22:490-493.

MAYR, E. 1969. Principles of Systematic Zoology. McGraw Hill, New York. x, 428pp.

MAYR, E., LINSLEY, E.G. & USINGER, R.L. 1953. Methods and Principles of Systematic Zoology. McGraw Hill, New York. ix, 336pp.

RIDE, W.D.L. 1982. Nomenclature of Organisms Treated Both as Plants and Animals. Biology International N°6:15-16.

RIDE, W.D.L. 1984. On the Organization and Financial Arrangements of Organizations Responsible for Botanical and Zoological Nomenclature. Taxon. 33:240-260.

RIDE, W.D.L. 1985. Introduction pp. xiii-xix in ICZN (1985). International Code for Zoological Nomenclature, 3rd Edn. (see above)

ROTHSCHILD, Lord. 1965. A Classification of Living Animals. Longmans, Green & Co., London. ix, 134pp.

STILES, C.W. 1980S. The International Code of Zoological Nomenclature as Applied to Medicine. Hygenic Laboratory Bull., N°24. Treasury Dept., Washington. 50pp.

STILES, C.W. 1926. International Rules of Zoological Nomenclature. Proc. Biol. Soc. Wash. 39:75-104.

STRICKLAND, H.E. et al. 1843. Series of Propositions for Rendering the Nomenclature of Zoology Uniform and Permanent, being the Report of a Committee for the Consideration of the Subject Appointed by the British Association for the Advancement of Scienc. Ann. Mag. Nat. Hist. (1) 11:259-275.

STOLL, N. 1961. Introduction. The International Code of Zoological Nomenclature. International Trust for Zoological Nomenclature, London.

THULBORN, T. 1986. Taxonomic Tangles from Australia. Nature. 321:13, 14.

TYLER, M.J. 1985. A Crisis in Zoological Nomenclature. Search. 16:237.

WAGENITZ, G. 1967. Betrachtungen uber die Artenzahlen der Pflanzen und Tiere. Sber. Ges. Naturf. Freunde (N.F.). Bd. 7:79-93.

WILEY, E.O. 1979. An Annotated Linnaean Hierarchy, with Comments on Natural Taxa and Competing Systems. Syst. Zool. 28:308-337.

NOMENCLATURE OF BACTERIA

P.H.A. Sneath[1]

HISTORY OF THE BACTERIOLOGICAL CODE

Origins of Bacteriological Nomenclature

The Bacteriological Code grew historically from the Botanical Code, which had been used for the nomenclature of bacteria in the early days of bacteriology. It is very close to the Botanical Code in its concepts and terminology. In the 1930's bacteriologists began to feel that the provisions of the Botanical Code did not fit too well the needs of bacterial systematics. At International Congresses of Microbiology in 1930, 1936 and 1939, there were moves to prepare separate Code for bacteria. Such a Code was approved at the 1947 Congress and first published in 1948 (Buchanan & Breed, 1948). Some revision and extension to include viruses, occurred in later years; the major changes were embodied in the International Code of Nomenclature of Bacteria and Viruses (1958). During this period the main figure was Professor R.E. Buchanan of Iowa State University, to whose vision and energy bacteriology has a special debt.

The Microbiology Congress in Moscow in 1966 marked a change of direction in the philosophy of bacterial nomenclature. The virologists decided to prepare their own rules of nomenclature. Bacteriologists began to feel that the Bacteriological Code had become too complex, and that the thousands of forgotten and useless names were a burden on systematics. Moves were set afoot that resulted in a central registry of names and a new starting date, 1980, for bacterial names. An account of these moves is given later; historical material will be found in the forewords to the 1958 and 1975 Codes and to the Approved Lists of Bacterial Names (Skerman et al., 1980).

Comparison of the Botanical and Bacteriological Codes

The Botanical and Bacteriological Codes have an almost uniform treatment of the form of names, of regular endings for rank categories, of typification (with one important exception); and of rules that operate on names of specified ranks. The excellent logical structure of the Botanical Code from which the Bacteriological Code was mostly developed has been a major advantage to bacteriologists, who have modified it very little in content, although the latest edition of the Bacteriological Code (International Code of Nomenclature of Bacteria, 1975) did involve a complete re-wording and re-arrangement of the rules themselves.

----

[1] Dr. Peter Sneath is a Member of the International Committee of Systematic Bacteriology and of its Judicial Commission; he was a Member of the Editorial Committee of the current Bacteriological Code.

Details of the differences between these two Codes may be found in the excellent exposition of Jeffrey (1973). The major differences are three.

1) The Botanical Code does not permit living types. In bacteriology these are encouraged, and are now mandatory for organisms that can be grown in pure culture.

2) The Bacteriological Code does not regulate the names of categories between subgenus and species (e.g., sections, series), nor names of hybrids or infrasubspecific forms (e.g., serological or pathogenic varieties).

3) The Bacteriological Code permits the conservation of names of any rank, and makes provision for exceptions to be made to any Rule by means of official Opinions.

The Bacteriological Code is responsible to International Congresses of Microbiology, through the International Committee for Systematic Bacteriology (ICSB). This committee consists of representatives of microbiological societies throughout the world, and coopts a limited number of other bacterial systematists. It has an Executive Board which sets up specialist taxonomic subcommittees to advise on taxonomy and nomenclature. These subcommittees are very active scientifically, and also liaise with bodies such as the World Health Organization. The main nomenclatural arm of the International Committee for Systematic Bacteriology is its Judicial Commission. The commissioners are elected by the ICSB from active bacterial systematists. The Commission is responsible for two major tasks, (1) preparing and amending the Bacteriological Code, and (2) adjudicating by Opinions on the Rules and on specific cases of nomenclature. The ICSB also has a Publications Committee which oversees publication of the Code and of its official journal, The International Journal of Systematic Bacteriology. The organization of the ICSB is described in an appendix to the Bacteriological Code (1975).

The Use of Living Types

The most important difference between the Bacteriological and Botanical Codes concerns living types. The majority of commonly-used properties of bacteria can only be determined by the activities of living cultures. The risks that type cultures may change with time, cultivation media, preservation methods or contamination, have to be weighed against the inadequacies of non-living type material. Bacteriologists have not found the disadvantages of type cultures to be serious, and they have been unable to work effectively without them.

The problems of living type material in biology may well have been significant in the past. Properties of living cultures do change with continuous cultivation, but this usually affects only a few properties such as loss of virulence of pathogenic strains: the cultures remain readily identifiable as to species. Very few bacterial type cultures have become unusable as type material, although such effects may be important over many decades or centuries. Technical advances, however, have provided solutions. Freeze-dried cultures remain viable and virtually unchanged for decades (though some plasmid-encoded genes may perhaps be lost because drying may adversely affect plasmids).   Cultures frozen at -196°C, the temperature of liquid nitrogen, probably will remain unchanged for many decades.

Further, the distinction between living and non-living types is becoming blurred. Freeze-dried DNA preparations would not be considered living, yet substantial lengths of genome may be recovered by genetic engineering; if sufficient genome is recovered the identity of the species that provided the freeze-dried DNA can be recognized unambiguously from its genes expressed in another living organism.   The recent success in recovering genes of the extinct quagga from dried museum tissue of this zebra (Higuchi et al., 1984) illustrates the point; the sequence of its cytochrome oxidase I is very similar to that of the mountain zebra. Conversely, dead preserved material from a type culture may well serve adequately as type material if serological or molecular techniques are available, although one could not produce sufficient material without first using what is in effect a living type culture.   A procedure to permit this in botany is now being explored (Friedmann & Borowitzka, 1982).

The essential point is therefore that the material, whether living or not, should permit unambiguous identification of the taxon of which it is the type, and the possibilities here have been greatly increased by recent chemical and molecular techniques.   The question of permanence (in the sense of the likelihood that specimens will remain intact, correctly labelled, and so on) is a secondary matter, reflecting curatorial practices.   The point of these arguments is especially relevant to the future taxonomy and nomenclature of microorganisms that we are only now learning to cultivate, such as blue-green algae and many protists.

A NEW START

The Problem Facing Bacterial Taxonomy

The new starting date and starting document for bacteria has been the most radical change in nomenclature in recent years (Anonymous, 1980).   It is therefore of some interest to record how this was achieved. Three concordant changes had to be made: (1) revision of the Bacteriological Code to give effect to the new rules; (2) compilation of the starting document of names that retained standing in nomenclature (the Approved Lists of Bacterial Names); and (3) establishment of an official publication or registry in which new names must be published to obtain validity.

The way in which bacteria were studied in the early years led to a great number of names that are today unrecognizable as to taxon. Bacteria were usually described on a few properties - morphological, cultural and physiological - they were given names but no type material was preserved, and registers of names were never effective. In consequence there were about 21,000 names of bacterial taxa in the literature according to the major compilation, Index Bergeyana (Buchanan et al., 1966). It was believed that many thousands of names had escaped notice, and a supplement (Gibbons et al., 1982) listed about 4,000 more names antedating 1966. Of almost 30,000 names, the vast majority were unrecognizable as to taxon (about 1,700 names with basonyms prior to 1966 reached the new starting document). Yet all had to be taken into account for synonymy and homonymy. This was at a time when numerous new bacterial species were being described. It was evident that a radical solution was needed.

At the 1966 Microbiology Congress it was decided to convene a special meeting of the Judicial Commission (see Sneath, this volume) to revise the Code (which was becoming cumbersome) and consider the problem of the thousands of useless names. The meeting was held in 1968, and Dr. S.P. Lapage, of the National Collection of Type Cultures, London, offered to undertake a complete revision of the Code. The Commission also agreed that valid publication should be restricted to a limited set of publications, and that for a name to be valid its nomenclatural type must be designated.

The Commission then went on to discuss the many useless names in the literature. The device used by the 1961 edition of the Zoological Code, whereby names disused for 50 years could be considered to be forgotten names (nomina oblita) and thereafter ignored, was not thought sufficient for the problem. I myself at first favoured the block conservation of well-established, recognizable, names in certain publications of international repute. Fortunately, the idea of preparing a new starting document was proposed by Professor V.B.D. Skerman of the University of Queensland. Though suggestions on new starting dates had been made in the past, the critical concept was this starting document, a modern Species Plantarum, as it were. Names not included would then lose their standing in nomenclature, and the thousands of forgotten names need no longer be considered.

It is a pleasure to pay tribute to Skerman's vision and persistence in advocating this solution over many years. In 1949 he had commented on the problem: "There is only one remedy for this, namely the redescription of all available cultures according to a certain code which should be applied to all bacteria alike. On the basis of these descriptions the organisms should be renamed, for the most part with the names they now possess. Priorities should be based on these names and all descriptions and names for which there are no procurable cultures should, by common consent, be discarded" (Skerman, 1949).

The development of new taxonomic techniques for bacteria soon resulted in considerable, though piecemeal, redescription of cultures on fairly consistent criteria (even if this was not as standardized as was envisaged in 1949). There was a great reduction in the number of names as it became evident that many names attached to cultures were synonyms. The trend is best shown by contrasting the numerous species listed in the 6th edition of Bergey's Manual of Determinative Bacteriology (Breed et al., 1957). In commenting on this progress Skerman (1968) said "a concerted effort should be made to ... recommend the conservation of names for which representative cultures are still extant or for which there is an unequivocal illustration or description on which to fix the type. These names could then be gazetted and a new date for recognizing priorities established". This last sentence represented the kernel of the matter. Formal conservation would be unnecessary if names predating the new starting date lost their standing in nomenclature the gazetted list of names would become the new starting document.

The Revised Code

Dr. Lapage made a thorough comparison of all the statements in the Bacteriological Code, and compared them with the corresponding statements in the Botanical and Zoological Codes. We are grateful to him for the care with which he considered the essential intent of each clause, and then redrafted them into simpler phraseology. He also devised a convenient and lucid set of infrasubspecific designations. He copied all the rules, recommendations and notes onto cards and rearranged them into a more logical order. Prof. Skerman provided an indispensable concordance between new and old versions. Aided by a small committee, Dr. Lapage then prepared a clear concise revision of the Code, taking particular care to distinguish obligatory rules from optional recommendations.

The major source of complexity was the decision to retain the categories of subgenus and subspecies. This led to retaining complex rules for their nomenclature, and also required the retention of autonyms; both of these are a constant source of irritation. Opinion was divided on the matter, and I must confess that at the time I believed that subgenus and subspecies would continue to fulfill a useful function. Since then I have come to the view that it would probably have been better to get rid of them, but only time will tell. We were, I think, uncertain whether the bacteriological community would be willing to acquiesce in their loss. Subspecies are occasionally still used, but subgenera are employed very rarely. A number of workers evidently require some device for entities near the species level whose best taxonomic treatment is disputed. We did get rid of variety (varietas) as we considered that either subspecies, or infrasubspecific categories like serovar, pathovar, would suffice. The infrasubspecific categories are not regulated by the new Code. It was recommended that the use of the term "type" should be restricted to nomenclatural types (e.g. neotype), so that serovar, for example, is now used instead of serotype.

There remained the provisions for the new starting date and document (these provisions were written in draft form then, but have since been given final wording (Rules Revision Committee, 1982), and also the introduction of a registry for all valid names. It was decided to implement the latter before the starting date, and from 1 January, 1976 all new names have had to be published in the International Journal of Systematic Bacteriology in order to be valid (further details are mentioned later). This provided a breathing-space before the publication of the starting document in 1980, in which valid names proposed between 1976 and 1980 were all found in a single journal; this greatly simplified the task of preparing the starting document.

A New Starting Document: the Approved Lists of Bacterial Names

The concept of the Approved Lists implicitly included the assumption that a good proportion of unnecessary synonyms had been identified. Though logically it would not have mattered if this had not been done, it would have made the preparation of the lists much more laborious, and would have left numerous nomenclatural problems on priority and the like. Considerable reduction in synonymy was achieved in all bacterial groups except the streptomycetes. These organisms had been studied intensively because of their importance in yielding antibiotics, and there were about a thousand names for strains that still were preserved in culture. The taxonomic revision needed to synonymize these could not be achieved in the available time (though some names could be discarded for other reasons).

The Approved Lists therefore contained over 400 names of species and subspecies of Streptomyces and Streptoverticillium, mainly based on an international collaborative survey of cultures (e.g., Shirling and Gottlieb, 1972 -Ed. have only cited the last of a series of papers). This represents over 20% of the species and subspecies names of bacteria in the Approved Lists, and the studies needed for synonymization are only now proceeding (e.g., Williams et al., 1983). Nevertheless, by insisting that the Approved Lists include type strains, these current studies are at least based on well-defined type material. They present only one major problem for nomenclature - whether the name of a given species should be that of the earliest original name before 1980 (as is required by the Code where priority of names within the Approved Lists has to be determined) - or whether a commonly-used name of later date should be conserved in the interests of stability.

Other groups of bacteria that would have presented the same problem as the streptomycetes were the genus Salmonella, causing food poisoning, and certain genera containing numerous named plant pathogens (e.g., Exanthomonas, Pseudomonas). However, it had by then been generally accepted that the numerous named forms of Salmonella were serological varieties of a few species, and the phytopathogenic forms were best treated as host-adapted pathogenic varieties. By treating these as infrasubspecific forms - serovars and pathovars respectively - they could be omitted from the Approved Lists. The pathovars, however, are mentioned again under re-use of names.

The main aim of the Approved Lists was to list the names of all those bacteria which (1) were clearly recognized to exist as separate species or subspecies, (2) for which a modern description existed, and (3) for which a type culture was available in a major culture collection. If noncultivable, the description or illustration was to serve as the type, as had been the case previously. The genera to which there was sufficient difference of taxonomic opinion to warrent listing the same species under two genera (e.g. Morganella morganii and Proteus morganii, with the same basonym Bacterium morganii and type strain ATCC 25830). The Lists, therefore, acknowledged that there could be legitimate differences of opinion on taxonomy, though fortunately there are few objective synonyms such as this.

For suprageneric names I suggested that those in a few major works should be listed unless they clearly contravened the rules, and these comprise the majority of the higher names.

Preparation of the Approved Lists

The preparation of the Approved Lists involved active participation of numerous individuals. The most important help came from the taxonomic subcommittees of the International Committee for Systematic Bacteriology, which reached most of the decisions on synonyms, on recognizability, and so on. Where no subcommittee was able to advise, the help of individual

systematists was enlisted. The operation was under an ad hoc committee of
the Judicial Commission; Prof. Skerman bore the brunt of the correspondence
and preparation, leading to the production at Queensland University of
camera-ready copy for publication in the International Journal of Systematic
Bacteriology. The bacteriological community also therefore owes its gratitude
to him for by far the greatest share in the work of preparing the Lists.

Many arbitrary decisions were taken by the advisors and compilers. It would
have been impossible to study the merits of all the nomenclatural problems
that arose. In general, well-established names were preferred over names that
might possibly have been synonyms with earlier priority: there would have
been no time to pursue such matters, and perhaps reach a conclusion that would
have later led to a request for conservation of the well-known name. Type
strains were chosen if necessary, and advisors were free to replace them by
lectotypes. The type species of a few genera were changed. In the event there
have been very few difficulties here. A later ruling by the Judicial
Commission (1985) made clear that types in the Approved Lists were to be
accepted as the correct nomenclatural types unless an appeal was lodged; such
appeals have been very few.

Nevertheless, the time to prepare the Approved Lists was uncomfortably
short, and a few errors of omission or commission crept in. There was not
time to double-check bibliographic citations, so a rather long list of minor
corrections was later published (Hill et al., 1984). It was, however, decided
that no additions or deletions of names should be made: the Approved Lists is a
fixed historical document; other names must be proposed and published in the
usual way. About 100 names that should have been included were
inadvertantly ommitted (mainly because they were published while the Lists
were being compiled), and these are having to be re-proposed.

The Approved Lists of Bacterial Names (Skerman et al., 1980) contains about
2,300 names, of which roughly 125 are suprageneric, 300 are generic, 1,800 are
specific and 100 are subspecific. The bulk, therefore, are species names
(nearly 80%), and the acceleration of taxonomic work on bacteria in recent
years is shown by the fact that about 40% of species were described in 1966-
1979 inclusive. A typical species entry may be cited as an illustration:

> Pediococcus dextrinicus (Coster & White, 1964) Back 1978.
> Coster, E. and H.R. White. 1964. Journal of General Microbiology
> 37:15-31; Back, W. 1978. International Journal of Systematic
> Bacteriology 28:523-527. Type strain: DSM 20335
> Description: Back, W. 1978. Ibid.

It should be noted that the entries are not intended to provide full details of
synonymy and the like. Thus the basonym is Pediococcus cerevisiae subsp.
Dextrinicus Coster and White, but the user can readily find this in the papers
cited. The modern description (with the type culture) fixes the application of
the name, even if the original description is poor (as is common with the
earliest literature). This description was common in the latest edition of the
standard monograph, Bergen's Manual (Buchanan & Gibbons, 1974).

A Registry of Names

The Judicial Commission was faced with the problem of how best to arrange an effective registry of new names and yet not restrict taxonomic freedom. All taxonomic papers in bacteriology could not be published in a single journal, nor would it be wise for one journal to have the power of censoring all taxonomic work. I suggested that this problem should be solved by exploiting the difference between effective publication and valid publication (in the sense of the Botanical and Bacteriological Codes), and this was accepted. Authors may publish names effectively in any scientific publication as before, but the names only become validly published when they are published in the official record. Effective publication requires a name in proper form together with a description. Valid publication requires also designation of the type (usually a type culture), unless done previously. If publication is in the official publication the name is both effectively and validly published.

To validate an effective publication the name, bibliographic references and nomenclatural type is "announced" in one of the lists that are issued periodically in the official registry, on receipt by the editor of evidence of effective publication (normally a reprint or xerox of page proofs). Priority dates from valid publication.

This procedure is unfamiliar to taxonomists, but it did not seem unreasonable to require authors to follow it if they wished their names to be afforded the protection of the Code. There has been some misunderstanding on this score: there is a belief that the registry should look for new names and validate them; or that priority should date from effective publication. A moment's thought will show these are impracticable. The ICSB has not the resources for the first. The latter would lead to endless problems when authors, after perhaps many years, claimed priority for names in obscure publications when they had never registered, and all the advantages of having to consult only one central registry would be lost.

At present, the only official publication for validity is the International Journal of Systematic Bacteriology, and lists of "announcements" are made several times a year. An example of such an announcement is given below (following the title "Validation of the Publication of New Names and New Combinations Previously Effectively Published Outside the IJSB", a List number, and an explanatory preface), published in April 1984.

Name	Proposed as	Authors	Nomenclatural Type
Clostridium lortetii	New species	Oren (10)	MD-2 (=ATCC 35059)

10. Oren, A. 1983. Clostridium lortetii sp. nov., a halophilic obligatory anaerobic bacterium producing endospores with attached gas vacuoles. Arch. Microbiol. 136:42-48.

The name has priority from the "announcement", April 1984. There are perhaps 30 such entries in one list (not to be confused with the Approved Lists).

No nomenclatural system is free from disadvantages. The reader can easily envisage things that may go wrong - incorrect citations or dates, delays in the mails, complaints that mistakes were not noticed - but the issue is whether this new system is any more prone to such problems than the old one. I believe one can say with some confidence after five years operation that the new system is working quite well. The official registry can easily accomodate the numbers of new names, running at about 120 a year. About 20 of these, however, are revived names, and we expect that the frequency of revival of names dating from before 1980 will soon fall off.

Re-use of Names

One important consequence of the new starting date and document is that all names published before 1980 lose their standing in nomenclature, unless included in the Approved Lists (when their standing dates from 1 January, 1980). One need not therefore search the pre-1980 literature. The converse, however, also holds: names before 1980, and those not yet validly published in the official registry, have no standing, and may be used freely. To demand otherwise would throw us back into lengthy literature searches. The essential problem is whether re-use of old names will cause confusion.

One new provision in the latest Code (1975) is to permit revival of names in the same sense as the original, for example because of rediscovery of a forgotten organism: such a revived name is attributed to the revivor, who gives a new description and type, but indicates its origin by citing the original author.

The re-use of a name in a sense other than the earlier name is a problem, because it may happen that the name had wide use in early literature, and re-use might be a source of confusion. No serious problems of this nature have yet arisen. The names of plant pathogenic varieties (pathovars) are frequently in latinized form, because they have in the past often been considered as distinct species. The International Society for Plant Pathology has therefore published a list of about 200 pathovar names, which bacteriologists are advised to avoid when creating new species names, and has suggested guidelines for future pathovar names (Dye et al., 1980). If serious confusion arises with re-used names the Judicial Commission has power to declare them rejected names (nomina rejicienda).

Timing of the Changes

The decision to make a new start in nomenclature was made by the Judicial Commission in 1968, which set up a committee to revise the Code. The Microbiology Congress in 1970 accepted the proposals in principle. The Code was revised in time for the 1973 Congress, which approved it and commissioned the Approved Lists. The revised Code was published late in 1975 and came into force on 1 January, 1976 (hence rather confusingly it is also called the Revised Code 1976: it contained the new requirement that valid publication must be in the offical journal and include designation of the type). A draft of the species names for the Approved Lists was published (Ad Hoc Committee, 1976, based largely on Buchanan & Gibbons, 1974)), and the advice of taxonomic subcommittees was requested. The Approved Lists were published on 1 January, 1980, and accepted at the 1982 Congress. Throughout this time a number of measures were taken to publicize the new proposals among bacteriologists, and to solicit advice and support for them.

THE FUTURE

New Starting Documents and Dates for Other Groups

A new start in nomenclature could be made in other groups of organisms, such as certain protista, microfungi or helminths.   Experience in bacteriology suggest that there are several conditions for success. First, the group must not be too large and there must be active workers willing to assist in the task. Second, taxonomists must accept that the philosophy must be that names lose their standing in nomenclature **unless there is good cause for retaining them;** this is the reverse of retaining them **unless they are removed for good cause.** It is not possible to achieve reforms if matters are continually held up in case something of historical value is lost; workers must accept that they are in the position of those who made the first starting documents in the eighteenth and ninetheenth centuries. Valuable material can later be added by revival if this is thought necessary. The main effect is on the loss of priority of a few names of historical interest.   Thirdly, there must be an effective central registry, which operates on liberal lines to accommodate differences of taxonomic opinion.

Minimal Standards of Description

One aim of the revisors of bacteriological nomenclature was to raise the scientific standards of taxonomic descriptions. To this end they incorporated recommendatoins in the new Code that for each group in turn there should be chosen a set of bacterial properties which all workers should endeavour to determine when preparing a description of a new taxon. These were then to be recommended to workers as minimal standards for taxon descriptions in that group. Some progress has been made, but the publication of minimal standards will be a continuing task for some time. Meanwhile, new techniques of study are introduced, so the standards will in due course need updating. However, the efforts to prepare such standards are having a benefical effect on the quality of systematic work.

Higher Categories in Procaryotes

In recent years, the study of ribonucleic acid has revolutionized the higher groupings of bacteria, due particularly to the work of Dr. C. Woese of the University of Illinois and his colleagues. It has become evident that most bacteria, "eubacteria", are genomically very similar to the blue-green algae (Cyanophytes or "cyanobacteria"). The latter can be considered one group of "eubacteria". A number of bacterial groups including the methanogens and extreme halophiles are much less similar to "eubacteria" than are the blue-green algae, and these have been referred to as "archaebacteria" (Woese, 1981). There has therefore been renewed interest in the highest groupings of the bacteria and other procaryotes (Murray, 1984). There is similar interest in higher groups in general (for example in protista, see Corliss, 1984).

From the point of view of nomenclature this development poses a particular problem - how to regulate the names of the highest taxa. The Botanical and Zoological Codes do not attempt to regulate names above the level of family and superfamily respectively. The Bacteriological Code carries this to the level of order, so that orders are based on the name of a type genus. It prescribes the latinized form of names above order, and mentions that types may be designated, but it does not require that priority is observed. There is evidently an uneasy transition in philosophy from the nomenclatural type concept at lower ranks, which may merit further study from nomenclaturalists.

The reasons for the difficulty at higher ranks seem to be principally because these are formed by division from above rather than agglomeration from below.   Consequently there is no assurance that higher groups are homogeneous: the invertebrates and the cryptogams are well-known examples. Several speakers at the nomenclature sessions of ICSEB III touched on this. In bacteria there are doubts that the "archaebacteria" are homogeneous.

It seems unlikely that these difficulties will be overcome by advances in technique, because they are inherent in groups produced by division from above. Division may indicate that organisms are different, but it does not show that the forms within a major division are similar to one another. Furthermore, numerical phylogenetic studies seem generally to show greatest uncertainty at higher ranks, and this is probably true of all phylogenetic work.

Organisms That Can be Treated Under More Than One Code

Problems with organisms that can be treated under more than one Code have arisen in bacteriology in two guises, homonyms within microorganisms, and the nomenclature of blue-green algae.

Homonyms Within Microorganisms

The Bacteriological Code has from its inception (1945 edition) forbidden the use for bacteria of homonyms of protista and similar microorganisms because of the risk of confusion. It therefore is committed in principle to suppress bacterial names that contravene this and replace them with new names, and such replacement has occurred in a number of times. Thus the bacterial genus Phytomonas was replaced by Xanthomonas because it was antedated by the protozoan genus Phytomonas (Dawson, 1939; Breed et al., 1957; International Code of Bacteriological Nomenclature, 1975). The International Committee for Systematic Bacteriology would therefore support current proposals to produce a checklist of genus names in protista and similar groups.

Nomenclature of Blue-Green Algae

The nomenclature of blue-green algae has also received a good deal of attention. Proposals have been made (Stanier et al., 1978) to treat these procaryotic organisms under the Bacteriological Code, but not all algologists have supported this.   The International Committee for Systematic Bacteriology has maintained close liaison with the officers of the Botanical

Code, and has sponsored several meetings on the topic, particularly at the XIIIth Microbiology Congress in 1982. It accepts the view of the botanists that workers are free to use whatever Code they believe is most appropriate but it wishes to see progress toward an agreed solution. It has recently set up a working party to study the matter in collaboration with botanists. It welcomes the progress being made by algologists toward acceptable procedures for type cultures and a new starting document (Friedmann & Borowitzka, 1982). The taxonomy of these organisms is in a state of transition, with at present few type cultures, and difficulties in relating cultured material with natural occurrences and ecology of these organisms.

Harmonization of the Codes

The harmonization of the Codes of Nomenclature remains a distant vision. Bacteriologists will however, welcome progress towards it, and the desiderata for an ideal Code presented by Jeffrey (see Jeffrey, this volume) will be broadly supported. It may be noted that nomenclature of living organisms is just about feasible with a binominal system: if harmonization becomes impossible, it may be worth considering an explicit trinominal system, with an additional name for the higher group involved. This is already commonly practiced de facto: many papers add the higher group name in parentheses. Although it would be a considerable break with tradition, this would alleviate pressure from homonyms and facilitate compilation of registers of names. Homonyms would then only be forbidden with each higher group (e.g., Classes, or perhaps Orders for insects), but it would be expected that the full tirnomen would be given at first mention.

REFERENCES

Ad Hoc Committee of the Judicial Commission of the ICSB. 1976. First Draft Approved Lists of Bacterial Names. Int. Jour. Syst. Bacteriol. 26:563-599.

ANONYMOUS. 1980. Debugging Systematic Bacteriology. Nature, London. 283:511.

BREED, R.S., MURRAY, E.G.D. & HITCHENS, A.P. 1948. Bergey's Manual of Determinative Bacteriology, 6th Ed. Williams & Wilkins, Baltimore. xvi + 1059pp.

BREED, R.S., MURRAY, E.G.D. & SMITH, N.R. 1957. Bergey's Manual of Determinative Bacteriology, 7th Ed. Williams & Wilkins, Baltimore. xviii + 1094pp.

BUCHANAN, R.E. & BREED, R.S. (eds.) 1948. International Bacteriological Code of Nomenclature. Jour. Bacteriol. 55:287-306.

BUCHANAN, R.E., HOLT, J.G. & LESSEL, E.F. (eds.) 1966. Index Bergeyana. Williams & Wilkins, Baltimore xiv + 1472pp.

BUCHANAN, R.E. & GIBBONS, N.R. (eds.) 1974. Bergen's Manual of Determinative Bacteriology, 8th ed. Williams & Wilkins, Baltimore. xxvi + 1246pp.

CORLISS, J.O. 1984. On the Systematic Position and Generic Names of the Gram Negative Bacterial Plant Pathogens. Zbl. Bakteriol. Hyg. Infectionsker. 2 Abt. 100:177-193.

DYE, D.W., BRADBURY, J.F., GOTO, M., HAYWARD, A.C., LELLIOTT, R.A. & SCHROTH, M.N. 1980. Standards for Naming Pathovars of Phytopathogenic Bacteria and a List of Pathovar Names and Pathotype Strains. Rev. Plant Pathol. 59:153-168.

FRIEDMANN, E.I. & BOROWITZKA, L.J. 1982. The Symposium on Taxonomic Concepts in Blue-Green Algae: Towards a Compromise with the Bacteriological Code? Taxon. 31:673-683.

GIBBONS, N.E., PATTEE, K.B. & HOLT, J.G. (eds.) 1982. Supplement to Index Bergeyana. Williams & Wilkins, Baltimore. vii + 442pp.

HIGUCHI, R., BOWMAN, B., FREIBERGER, M., RYDER, O.A. & WILSON, A.C. 1984. DNA Sequences from the Quagga, An Extinct Member of the Horse Family. Nature, London. 312:282-284.

HILL, L.R., SKERMAN, V.B.D. & SNEATH, P.H.A. (eds.) 1984. Corrigenda to the Approved Lists of Bacterial Names. Int. Jour. Syst. Bacteriol. 34:508-511.

International Code of Nomenclature of Bacteria and Viruses. 1958. Edited by the Editorial Board of the International Committee on Bacteriological Nomenclature. Iowa State College Press, Ames, Iowa. xxii + 186pp.

International Code of Nomenclature of Bacteria. 1975. S.P. Lapage, P.H.A. Sneath, E.F. Lessel, V.B.D. Skerman, H.P.R. Seeliger & W.A. Clark, eds. American Society for Microbiology, Washington, D.C. xxxv + 180pp.

Judicial Commission. 1985. Opinion Nº 58. Confirmation of the Types in the Approved Lists as Nomenclatural Types Including the Recognition of Nocardia asteroides (Eppinger, 1891), Blanchard, 1896 and Pasteurella multocida (Lehmann & Neumann, 1899), Rosenbusch & Marchant, 1939, as the Respective Type Species of the Genera Nocardia and Pasteurella and Rejection of the Species Name Pasteurella gallicida (Burrill, 1883), Buchanan, 1925. Int. Jour. Syst. Bacteriol. Volume 35.:538

JEFFREY, C. 1973. Biological Nomenclature. Edward Arnold, London. ix + 69pp.

MURRAY, R.G.E. 1984. The Higher Taxa, or, a Place for Everything. In N.R. Krieg & J.G. Holt (eds.). Bergey's Manual of Systematic Bacteriology, Vol. 1, pp. 31-34. Williams & Wilkins, Baltimore. xxvii + 964pp.

Rules Revision Committee. 1982. Proposals to Amend the International Code of Nomenclature of Bacteria. Int. Jour. Syst. Bacteriol. 32:142-143.

SHIRLING, E.B. & GOTTLIEB, D. 1972. Cooperative Descriptions of Type Strains of Streptomyces. V. Additional Descriptions. Int. Jour. Syst. Bacteriol. 22:265-394.

SKERMAN, V.B.D. 1949. A Mechanical Key for the Generic Identification of Bacteria. Bacteriol. Rev. 13:175-188.

SKERMAN, V.B.D. 1968. The Status and Scope of the Taxonomy of Bacteria. pp. 1-20. In: Taxonomy of Microorganism: Proc. 10th Institute of Applied Microbiology Symposium. University of Tokyo, Tokyo.

SKERMAN, V.B.D., McGOWAN, V. & SNEATH, P.H.A., (eds.) 1980. Approved Lists of Bacterial Names. Int. Jour. Syst. Bacteriol. 30:225-420.

STANIER, R.H. SISTROM, W.R., HANSEN, T.A., WHITTON, B.A., CASTENHOLZ, R.W., PFENNIG, N., GORLENKO, V.N., KONDRATIEVA, E.N., EIMHJELLEN, K.E., WHITTENBURY, R., GHERNA, R.L. & TRUPER, H.G. 1978. Proposal to Place the Nomenclature of the Cyanobacteria (Blue-Green Algae) Under the Rules of the International Code of Nomenclature of Bacteria. Int. Jour. Syst. Bacteriol. 28:335-336.

WILLIAMS, S.T., GOODFELLOW, M., ALDERSON, G., WELLINGTON, E.M.H., SNEATH, P.H.A. & SACKIN, M.J. 1983. Numerical Classification of Streptomyces and Related Genera. Jour. Gen. Microbiol. 129:1743-1813.

WOESE, C.R. 1981. Archaebacteria. Scientific American. 244:98-122.

NOMENCLATURE OF PLANT VIRUSES

David W. Kingsbury[1]

Introduction

Classification and nomenclature are logically separable intellectual processes. This has led to the development of official codes of nomenclature and a need for specialists in that activity within the venerable disciplines of zoology and botany. But those disciplines have long histories because they deal with objects visible to the naked eye, whereas recognition of viruses as a distinctive class of invisible organisms is a product of investigations begun less than a century ago. Indeed, only the last few decades mark the development of a clear understanding that viruese are obligate intracellular parasites comprising only a few genes in the form of either DNA or RNA. Thus, it is understandable that the taxonomy of viruses is a developing discipline, with very few virologists concerning themselves exclusively with taxonomic scholarship (as distinguished from identification, practiced by many clinical virologists), and that virus nomenclature has not yet been split off as a separate intellectual activity.

A Code Of Virus Nomenclature

The International Committee on Taxonomy of Viruses (ICTV), which has become the taxonomic authority of the community of virologists by their mutual consent, has so far generated only a rudimentary "code of nomenclature". This is embedded along with several taxonomic prescriptions in a set of Rules and Guidelines for distinguishing virus species (Table 1). Some of these rules may shock workers in other fields. Rule 4 does not mandate a latinized nomenclature, but only urges that "an effort be made" to conceive one. Although rule 5 would save any Latin species names already in use, there are few, if any, to save. The death of Latin is also obliquely acknowledged by the admonition that study groups recognize national sensitivities with regard to language (guideline 5). Additional permissiveness is displayed by the admission of sigla (rule 7), hyphenated words, appended numbers or letters (rules 14 and 15), and unwillingness to outlaw subscripts, superscripts, hyphens, oblique bars or Greek letters (guideline 4).

The permissiveness of the International Committee on Taxonomy of Viruses is clearly an effort to reconcile opposing viewpoints. Matthews and his contributing authors (Matthews, 1983) have reviewed the history of these controversies, which need not be repeated here. Suffice it to say that from the inception of organized efforts to classify viruses several decades ago, there has been adamantine resistance to a binomial latinized nomenclature by several vocal virologists among thoses who specialize in pathogens of plants (Milne, 1984; Murant, 1985).

----

[1] Dr. David Kingsbury is a Member of the International Committee on Taxonomy of Viruses, and of the Vertebrate Virus Sub-Committee.

Philisophical Justifications

This opposition to a traditional style of nomenclature is ostensibly based on two philosophical arguments about the essence of viruses. The first argument is that viruses are not "organisms", because of their extreme genetic simplicity. Rather, they deserve to be classified as a kind of subcellular molecular assemblage. The second argument is that the classical concept of biological species (Mayr, 1982) is inapplicable to viruses, because it is defined in terms of sexual reproductive compatibility, and viruses reproduce asexually.

Both arguments are easily rebutted. The one concerned with genetic simplicity is fallacious, because it equates a difference in degree with a difference in kind. All life is composed of chemicals. It is arbitrary to exclude viruses from biology on grounds that they have less genetic material and fewer genes than other organisms, since a continuum of genetic complexity exists. I submit that any extrachromosomal entity possessing genetic material and capable of replication, even if not autonomously, is a bona fide biological organism. Bacterial plasmids and eukaryotic episomes have levels of genetic complexity comparable to the simpler viruses. Would we exclude these entities from biology along with the recombinant DNA technology that exploits them and the information about the evolution and behavior of genes from higher organisms that this technology is delivering?

The second argument fails because it rests on a principle that was conceived by biologists who work with organisms that reproduce sexually, to satisfy their perceived need for a monistic criterion to distinguish species within their disciplines. In fact, asexual reproduction is known in zoology and botany, so this criterion is not universally applicable even in those fields, and bacteriologists have been productively classifying and naming myriads of asexually reproducing microbial species for a century. Therefore, despite its parochial utility for some taxonomists, the limited applicability of the criterion of sexual compatibility reveals that it is an epiphenomenon and that reliance upon it actually evades the issue of what constitutes a species. A more universal and fundamental definition would emphasize the concept that a species is a population of organisms sharing a pool of genes that is normally maintained distinct from the gene pools of other organisms, without stipulating the mechanism responsible for maintaining that distinctness. For asexual organisms, a major speciating force is ecology (Mayr, 1982). As applied to viruses, which are parasites, ecology includes host range, mode of transmission, and tissue tropisms. Considering the variety of life styles exhibited by organisms, other mechanisms of speciation independent of sexuality are also likely.

The Real Issues

The weaknesses of these philosophical arguments about taxonomy suggest that they were conceived post hoc to rationalize an aversion to Latin names that really came first (also see Ride, this volume). That aversion is also shared by many virologists who work with pathogens of organisms other than plants, contributing to the readiness of the International Committee on Taxonomy of Viruses to accept a loosening of the reins in its nomenclatural rules and guidelines (Table 1). This should have brought all virologists together. However, having constructed a philosophical edifice that precludes the application of the concept of biological species to viruses, the vocal plant virologists are having difficulty abandoning their position and accepting the relaxation of nomenclature offered by the Committee. So nomenclature, the servant of taxonomy, having mistakenly been made its master, created needless dissension and confusion that may continue to be difficult to resolve (Matthews, 1983; Milne, 1984; Murant, 1985).

In the meantime, under the auspices of the Committee, two taxonomic systems coexist. One, embraced by the plant virologists, eschews the traditional designations of taxa that the rest of the Committee's membership accepts. However, as shown in Table II, which compares the two systems, they are really equivalent conceptually, the same hierarchical levels being designated by different terms. Thanks to the flexibility of the Committee and the continuing research of plant virologists, plant virus classification is actually advancing steadily and logically along lines parallel to those of tradition, despite the controversies I have described. Fortunately, therefore, there are unlikely to be any long-term practical disadvantages if the present philosophical/semantic dichotomy continues to be unresolved. It also seems probable that the break with tradition represented by the acceptance of a non-Latin species nomenclature will not diminish the self-respect of virologists or the reputation of virology as a biological science.

Acknowledgements

The author is supported by Cancer Center Grant CA 21765 from the National Cancer Institute and by American Lebanese Syrian Associated Charities of St. Jude Children's Research Hospital.

Table 1. The Rules of Nomenclature of Viruses[a]

RULES

1. The International Code of Nomenclature of Bacteria shall not be applied to viruses.

2. Nomenclature shall be international.

3. Nomenclature shall be universally applied to all viruses.

4. An effort will be made towards a latinized nomenclature.

5. Existing latinized names shall be retained whenever feasible.

6. The Law of Priority shall not be observed.

7. Sigla may be accepted as names of viruses or virus groups, provided that they are meaningful to workers in the field and are recommended by international virus study groups.

8. No person's name shall be used.

9. Names should have international meaning.

10. The rules of orthography and epithets are listed in chapter 3, section 6 of the proposed International Code of Nomenclature of Viruses (Appendix C; Minutes of 1966 (Moscow) meeting).

11. A virus species is a concept that will ordinarily be represented by a cluster of strains, or a population of strains from a particular source, which have in common a set or pattern of correlating stable properties that separates the cluster from other clusters of strains.

12. The genus name and species epithet, together with the strain designation, must give an unambiguous identification of the virus.

13. The species epithet must follow the genus name and be placed before the designation of strain, variant, or serotype.

14. A species epithet should consist of a single word or, if essential, a hyphenated word. The word may be followed by numbers or letters.

15. Numbers, letters or combinations thereof may be used as an official species epithet where such numbers or letters already have wide usage for a particular virus.

16. Newly designated serial numbers, letters or combinations thereof are not acceptable alone as species epithets.

17. Artificially created laboratory hybrids between different viruses will not be given taxonomic consideration.

18. Approval by the International Committee on Taxonomy of Viruses (ICTV) of newly proposed species, species names and type species will proceed in two stages. In the first stage, provisional approval may be given. Provisionally approved proposals will be published in an ICTV report. In the second stage, after a 3-year waiting period, the proposals may receive the definitive approval of the ICTV.

19. The genus is a group of species sharing certain common characters.

20. The ending of the name of viral genus is "...virus".

21. A family is a group fo genera with common characters, and the ending of the name of a viral family is "...viridae".

22. Approval of a new family must be linked to approval of a type genus; approval of a new genus must be linked to approval of a type species.[b]

GUIDELINES

1. Criteria for delineating species may vary in different families of viruses.

2. Wherever possible, duplication of an already approved species name should be avoided.

3. When a change in the type species is desirable, this should be put forward to ICTV in the standard format for a taxonomic proposal.

4. Subscripts, hyphens, oblique bars, or Greek letters should be avoided in future virus nomenclature.

5. When designating new virus names, study groups should recognize national sensitivities with regard to language.

6. ICTV is not concerned with the classification and naming of strains, variants or serotypes. This is the responsibility of specilist groups.

7. Virus taxonomy at its present stage has not evolutionary or phylogenetic implications.[c]

- - - - - - - - - - - - - - - - - -

[a] - From (Matthews, 1981)

[b] - The concept of type genus was recinded by the Executive Committee of the ICTV in May, 1983.

[c] - Guideline 7 was deleted by the ICTV Executive Committee in May, 1983.

Table II. Coexisting Taxonomic Systems for Viruses*

Traditional	Plant Viruses
Family	Group
Genus	Subgroup
Species	Type (or Virus)

*adapted from (Milne, 1984)

REFERENCES

MATTHEWS, R.E.F. 1981. The Classification and Nomenclature of Viruses: Summary of Results of Meetings of the International Committee on Taxonomy of Viruses in Strasbourg, 1981. Intervirology. 16:53-60.

MATTHEWS, R.E.F. (ed.) 1983. A Critical Appraisal of Virus Taxonomy. CRC Press, Boca Raton, Florida.

MAYR, E.W. 1982. The Growth of Biological Thought: Diversity, Evolution and Inheritance. Harvard University Press, Cambridge, Massachusetts.

MILNE, R.G. 1984. The Species problem in Plant Virology. Microbiol. Sciences. 1:113-117.

MURANT, A.F. 1985. Taxonomy and Nomenclature of Viruses. Microbiol. Sciences. 2:218-220.

ASPECTS OF THE NOMENCLATURE OF FUNGI

Paul M. Kirk
and
David L. Hawksworth[1]

Starting Point Date

The 1910 (Brussels) International Botanical Congress adopted three different starting point works for living, as opposed to extinct, fungi. These were Linnaeus's Species Plantarum for myxomycetes and lichenized fungi, Persoon's Synopsis methodica Fungorum for rusts, smuts and gasteromycetes, and Fries' Systema mycologicum for all the other fungi. At the 1950 (Stockholm) International Botanical Congress these three starting point works were replaced by the dates at which they were considered to have appeared: 1 May, 1753, 31 December, 1801 and 1 January, 1821 (vol. 1 only) respectively. Names introduced before these dates were "devalidated" and their valid publication was displaced to their first post-starting point use. As such, any post-starting point publication, taxonomic or otherwise, acquired a nomenclatural importance never intended by its author.

This system of later starting points necessitated time consuming bibliographical searches to determine where a pre-starting point name was validated and caused considerable instability in names and citations as earlier works were repeatedly discovered (see Demoulin et al., 1981). To overcome this problem, the deletion, from Art. 13.1, of later starting points for fungi was effected at the 1981 (Sydney) International Botanical Congress following proposals made by the International Mycological Association (IUBS Section for General Mycology). However, the application of later starting points for more than 70 years has necessitated, at least temporarily, the granting of a protected status to names which were used in the former starting point works; such names are said to be sanctioned and are treated as if conserved against all earlier homonyms and synonyms. However, since priority is always rank-limited, sanctioning is also dependent on rank; sanctioned names are consequently protected only at the rank assigned to them by their sanctioning author.

In order to indicate that a name is sanctioned, the sanctioning author's name may, if desirable, be placed after that of the publishing author and separated by a colon. The appearance of a colon in an author citation does not, however, indicate the rank at which the name or epithet was sanctioned and should only be treated as an indication that the name or epithet has special nomenclatural status. With such an author citation we have, therefore, an indication of who

----

[1] Dr. David Hawksworth is Chairman of the Special Committee for Fungi of the International Association for Plant Taxonomy and Chairman of the International Commission of the Taxonomy of Fungi of the International Union of Microbiological Societies; Dr. Paul Kirk is Editor of the Index of Fungi.

has validly published the name and thus a clue as to where and when it was published, and also information on the nomenclatural status of the name. For example, the name Agaricus campestris was introduced by Linnaeus and subsequently sanctioned by Fries and can be cited as Agaricus campestris L. or Agaricus campestris L.: Fr.

The inclusion in an author citation of an element which is nothing to do with the valid publication of the name, is however, a concept foreign to the rest of biological nomenclature and in our view is against the trend to greater unification in biological nomenclature as a whole. We suggest that it should be used only as a form of short-hand in nomenclatural discussions. The matter is currently being addressed by the Special Committee for Fungi of the International Association for Plant Taxonomy.

The change of starting point also removed an instance where a biological judgement had to be made before the Botanical Code could be applied: the question of lichenization. Prior to 1981, the lichenized and non-lichenized fungi had starting point dates of 1753 and 1821 respectively. In cases where it was uncertain whether a fungus was lichenized or not, or where lichenization may either be facultative or occur at particular stages in development, and bearing in mind that the concept of precisely what fungus-alga associations should be regarded as "lichens" is subjective, it was effectively impossible to arrive at a satisfactory decision as to which starting point to adopt. Also, since lichenization is a biological phenomenon, and not necessarily of fundamental taxonomic importance, at least 20 genera exist which contain both lichenized and non-lichenized species (Hawksworth, 1978). In these cases species within the same genus had different starting points where their nutritional method varied. There was also the case where lichenized species were described in genera in which the type was non-lichenized. The genus would date from 1821 although the lichenized species were, if Art. 43.1 was ignored (as has been the usual practice of lichenologists), validly published under the same generic name earlier. There still exists, however, some uncertainty as to whether sanctioning should or should not cover the names of lichenized fungi and this will require resolution at the Berlin International Botanical Congress.

With the change to a uniform 1753 starting point the above anomalous situations were removed. However, there still exists a potentially more serious problem for we still have to make a taxonomic decision in certain borderline cases as to which of the two sanctioning works, those of Fries (1 Jan., 1821) and Persoon (31 Dec., 1801), we follow. One example here will suffice.

The second largest of the major groups of fungi is the Basidiomycotina. Here are classified the familiar mushrooms and toadstools and their allies. One of the major divisions in this group is made on the presence or absence of forcibly discharged basidiospores. Thus, the Hymenomycetes are distinguished by ballistosporic basidiospores and the Gasteromycetes by statismosporic basidiospores. Here lies the problem, for we follow one sanctioning work (Fries) for the Hymenomycetes and the other (Persoon) for the Gasteromycetes. There exists a number of well known groups of basidiomycetes which are probably monophyletic but which include genera containing species which are ballistosporic and other genera with species characterized by statismosporic basidiospores. It is clearly unsatisfactory that both sanctioning works should apply to such undoubtedly closely related groups of fungi.

The solution to the problem of the application of early names adopted by the bacteriologists, that is, the adoption of a list of approved names, has much to commend it. At present this is not a practical proposition with the fungi due to the large numbers involved, currently over 65,000 accepted species and perhaps as many as 180,000 names. However, in the event of International Registration Procedures being developed for newly described taxa, the possibility of preparing such lists should be re-investigated.

Pleomorphic Fungi

The second key feature of the International Code of Botanical Nomenclature applicable only to the fungi is that a single "species" can have more than one correct name, that is both validly published and legitimate. This feature concerns species with pleomorphic life-cycles. The stage in the life-cycle characterized by meiotic spore formation is termed the "teleomorph" and the name applied to this stage is also the name of the "holomorph", the species in all its morphs. The stage in the life-cycle characterized by mitotic spore formation is termed the "anamorph"; if more than one type of mitotic form is found, termed "synanamorphs", it is possible to give additional names to these morphs, if it is thought necessary or desirable.

An example of a fungus of this type is Thielaviopsis basicola (Berk. & Broome) Ferraris, the causal agent of a black root rot of tobacco and other plants. Here the Thielaviopsis morph is that which is usually encountered on the natural substratum. However, when the fungus is isolated in pure culture and grown on an agar medium a morph referable to a different genus is produced and this is named Chalara elegans Nag Raj & W. Kendrick. A similar fungus, Thielaviopsis paradoxa (de Seynes) von Hohnel, causing disease on a variety of plants such as pineapple, oil palm, banana, sugar cane and cocoa, also has a Chalara synanamorph, Chalara paradoxa (de Seynes) Sacc. Here, however, both synanamorphs have been correlated with their teleomorph, which is named Ceratocystis paradoxa (Dade) C. Moreau. In such cases, where the teleomorph is rarely found in nature, it is convenient to refer to collections of the fungus as, for example, the Thielaviopsis state of Ceratocystis paradoxa. This is especially important in applied mycology where it may be the anamorph which is the invasive plant pathogen or producer of particular enzymes or secondary metabolites.

The Problem of Kingdoms

In systems of classification of living organisms problems frequently arise when attempting to define unambiguous boundaries. The living world clearly cannot be neatly divided into a small number of monophyletic kingdoms because of the enormous diversity of organisms. In addition, more sensitive techniques are being actively developed at this time, particularly with reference to the protistan groups and results from the application of these techniques will undoubtedly throw further light on phylogeny. Understanding phylogeny is basic to all natural systems of classification. In the case of the fungi, however, the fossil record is very poorly researched, at least by non-palaeontologists, and this lack of solid information has permitted a wide variety of opinions especially concerning organisms which do not conveniently fit into traditional phylogenies.

The slime moulds, the Myxomycota or Mycetozoa, and their allied, combining an animal-like feeding stage and a plant-like fruiting stage, have long intrigued both botanists and zoologists. This has resulted in their classification in both the animal and plant kingdoms and, therefore, the use of two Codes, with implications for the endings of family names and the criteria for valid publication.

In 1858, an early student of these organisms, the mycologist de Bary, considered them distinct from the plants and believed they had evolved independently of the fungi. The new kingdom Protista was introduced by Haeckel and the slime moulds were included here. This placement was accepted by Whittaker (1969) who introduced the five-kingdom system comprising Monera, Protista, Fungi, Plantae and Animalia.

There is now universal acceptance of the Fungi as a separate Kingdom (Hawksworth, Sutton & Ainsworth, 1983). The typically mycelial trophic stage of the fungi with non-photosynthetic absorptive nutrition adequately distinguishes these organisms from those of other groups. Although the slime moulds and their allied are often included in the fungi their phagotrophic mode of nutrition, type of flagellation, amoeboid movement, localization of meiosis in developing spores and important biochemical differences have been cited as reasons why it is highly unlikely that they are related to any groups of the true fungi. Further, the fungi with flagellate zoospore stages (Oomycetes) clearly have protistan affinities and are increasingly segregated by protistologists; their isolation is also supported by biochemical, ultrastructural and physiological characteristics.

The problem, therefore, is that logically the slime moulds should be treated under the Zoological Code yet traditionally they have been studied by mycologists and treated under the Botanical Code. A change would therefore be very unstabalizing and so the majority of workers in this field use the Botanical Code although the other code is available for application to these organisms by those who consider them to be animals.

However, we see no problem in workers on different Kingdoms "adopting" a Code purely as a matter of preference for nomenclatural purposes bearing in mind that this is already the situation with the Fungi and Cyanobacteria. Similarly, we would hope that the Oomycetes can continue to be governed with the other Fungi under the Botanical Code; if in the future there were to be a separate code for protistan names and the Oomycetes were to be excluded from the Botanical Code, many unfortunate changes in the names of organisms of economic importance could result.

REFERENCES

BARY, A. de. 1858. Ueber die Myxomyceten. Bot. Zeitung. 16:357-358, 361-364, 365-369.

DEMOULIN, V., HAWKSWORTH, D.L., KORF, R.P. & POUZAR, Z. 1981. A Solution of the Starting Point Problem in the Nomenclature of Fungi. Taxon. 30:52-63.

HAWKSWORTH, D.L. 1978. The Taxonmy of Lichen-Forming Fungi: Reflections on Some Fundamental Problems. In Essays in Plant Taxonomy (H.E. Street, Ed.). pp.211-243. Academic Press, London, etc.

HAWKSWORTH, D.L., SUTTON, B.C. & AINSWORTH, G.C. 1983. Ainsworth & Bisby's Dictionary of the Fungi. 7th Edition. Commonwealth Mycological Institute, Kew.

WHITTAKER, R.H. 1969. New Concepts of Kingdoms of Organisms. Science. 163:150-160.

SOME ASPECTS OF ZOOLOGICAL NOMENCLATURE

The differences between the various codes of biological nomenclature have been the subject of increased attention in recent years. The depth and breadth of those differences have thus been revealed, at least to an extent sufficient to show the price of implementing measures of unification of the codes. This price would be changes on a very wide scale in the names of familiar organisms. Name changing of that order of magnitude would be directly contrary to the main purpose of all the codes, namely, stability of nomenclature.

The most important differences are those between the zoological and the botanical codes, for as the accompanying articles show, bacteriology and virology have a compactness that allows their nomenclatures to be controlled in largely empirical ways. The differences in question are clearly set out by Jeffrey (see Jeffrey, this volume). Two of the Features that differentiate zoological from botanical nomenclature are of particular significance. These are the principle of coordination (Jeffrey's point 4) and the notion of the nominal taxon (Jeffrey's point 10). I propose to consider briefly the role of these features in the present zoological code and to explain why they are important to zoologists.

The Principle of Coordination

The principle of coordination is enunciated in Articles 36, 43 and 46 of the third edition of the International Code of Zoological Nomenclature (1985) and in the Glossary. It states that "within the family group, or the genus group, or the species group, a name established for a taxon at any rank within the group is deemed to be simultaneously established with the same author and date for taxa based on the same name-bearing type at other ranks in the group". The logic behind the principle is simple: since the decision to allot specific or subspecific or generic or subgeneric rank to a new taxon embodies a subjective element (as also does any subsequent decision as to its ranking), it is absurd to pretend that there is any objective difference between a nominal taxon at any rank in a given group and another taxon based on the same name-bearing type at any other rank in the same group. The establishment of one implies the simultaneous (though logically consequent) establishment of the other.

The central core that gives the principle of coordination its strength is the name-bearing type. Zoological nomenclature does not limit the number of ranks that may be recognized in the family group (or now, to all intents and purposes, in the genus group); but if a series of taxa exists, one at each rank in

----

[1] Mr. Richard Melville was Secretary of the International Commission on Zoological Nomenclature from 1968-1985. This paper formed Mr. Melville's introduction to the symposium "Codes of Nomenclature" that provided the basis for the present volume.

the group, all based on the same name-bearing type, then they are all coordinate; that is, they all bear the same name (except for the suffix in family-group names) and all have the same author and date. A valuable simplicity and flexibility are thus imparted to the nomenclature without damage to stability.

The Nominal Taxon

The nominal taxon is an important element in zoological nomenclature, since it is the means whereby zoologists distinguish between the thing named and the mere name. It is defined in the Glossary to the Code as a "nomenclatural concept denoted by an available name and based, actually or potentially, on its name-bearing type, but having no defined taxonomic boundaries".

Botanical nomenclature speaks of "the type of a name"; zoological nomenclature of the type of the thing named. What is named is a concept that is subject to any number of different interpretations, and that can only be made operationally useful if all the interpretations have a common point of reference. This point of reference is the type of the concept, that which bears the name o the concept, and that without which the name must never be used.

Zoological nomenclature sees a logical fallacy in the notion of the type of a name, for that treats the name as being the same as the thing named, which is manifestly not the case. In taxonomic communications, a particular name may at one time be generally understood as referring to only one named concept based on only one name-bearing type; at a later date, it may be generally understood as being the valid name of n concepts, each based on its own name-bearing type, and each retaining enough viability to be resuscitated as a self-standing taxon at any time. Indeed, at a given time, "n" may not have the same value for two interlocutors, but the device of the nominal taxon, or the nominal taxa involved, can help them to identify the area of agreement between them, and thus to communicate constructively.

One problem area in which zoologists find the nominal taxon particularly useful is one that other biologists have difficulty in understanding. This is in recognizing and dealing with misidentifications. These do not seem to occur with anything like the same frequency in other areas of biology as in zoology, where they present difficulties in every aspect of taxonomic work, but especially in identification services to applied zoology and palaeontology. The difficulty they present is inherent in the subjective nature of all taxonomic work. Yet the correct identification of individuals to their species demands a thorough knowledge of the nominal species (singular or plural) involved and of its or their types. It is in the mental balancing of the concepts represented that the art of the zootaxonomist largely resides.

Another area where the nominal taxon imparts a particular stamp to zoological nomenclature is in homonymy, as Jeffrey shows (see Jeffrey, this volume). Once a specific name ("epithet" in botany) has been made available in combination with a generic name, a nominal species has been created. It can be recognized by that specific name (tied to its type) wherever it may go from one genus to another. This is not more than to recognize that the moving thing is not merely the name, but the concept labelled by the name, and that concept is immortal.

Zoological nomenclature is not only the servant of zoological taxonomy, it is also the vocabulary of the language in which zoologists communicate. The device of the nominal taxon may seem to reduce the importance given to names as such in relation to their meanings. This surely puts things in the right proportions. The baggage of meaning that a name carries with it is of far more importance than the name itself. The rules of nomenclature merely assist clarity of communication by ensuring that the name remains the same through taxonomic vicissitudes of the species. They must also provide a framework that ensures the permanence of the concepts that have been named.

SOME DIFFERENCES BETWEEN THE BOTANICAL AND ZOOLOGICAL CODES

C. Jeffrey[1]

The purpose of a Code of Nomenclature is to regulate nomenclature to ensure that, with any given circumscription, position and rank, a taxon can have only one name by which it may properly be known. In the historical development of biological nomenclature, this end has been found to be best served by the acceptance of three basic operative criteria - publication, typification and priority. The requirements of publication ensure the adequate documentation and dating of names, of typification, the fixation of the application of names to taxa, and of priority, a means of deciding the name to be used when the types of two or more names (or nominal taxa) fall within the range of variation of one (taxonomic) taxon.

The acceptance of these requirements is common to both the Botanical and the Zoological Codes. Independence of development since the time of Linnaeus has, however, resulted in major differences in their detailed application between the two Codes. These differences have two important consequences. First, they are inconvenient and confusing for workers whose study may be, or have been, referred to both plant and animal kingdoms and which may therefore have different names by which they should properly be known, according to which Code is followed; secondly, they form a massive obstacle in the way of any attempt to unify the two Codes. In view of these consequences, a brief review of some of the differences that might be considered important may be found useful, and is therefore presented here. In the following discussion, B denotes Voss, E.G., et al., International Code of Botanical Nomenclature, adopted by the Thirteenth International Botanical Congress, Sydney, August 1981 (1983), and Z denotes International Commission on Zoological Nomenclature, International Code of Zoological Nomenclature, Third Edition, adopted by the XXIInd General Assembly of the International Union of Biological Sciences (1985); unless otherwise indicated, the numbers quoted after these abbreviations refer to the Articles of the respective Code.

1. The Codes assert their mutual independence (B Principle I; Z 1c). Thus Pieris brassicae (a butterfly) can with perfect nomenclatural propriety pollinate Pieris ovalifolia (an ericaceous shrub); more seriously, a name that may be properly used for a taxon under one Code may ba a later homonym under the other. They also differ in scope, the Zoological Code, unlike the Botanical, not regulating names above the suprafamilial nor below the subspecific levels (B 3.1., 4, 16-27; Z 1b 4-5, 10c, 16, 45 f-g).

2. Different criteria for valid publication (B 32-35) or availability (Z 10-20) are specified by the two Codes; for example, Latin diagnoses are generally required as a condition of valid publication in botany (B 36), but are not needed in zoology (Z 12, 13); names of families and other higher taxa published with non-Latin terminations are not to be accepted (B 18.4) in botany, but may be

----

[1] Mr. Charles Jeffrey is a Principal Scientific Officer at the Royal Botanic Gardens.

made available in zoology (Z 11 f iii); and a species name cannot be validly published (B 43), but in zoology, providing the specific name is included in a binomen, the generic name need not even be available (Z 11 h iii 1). The names of taxa (other than of algae) originally assigned to a group not covered by the Botanical Code when treated as plants have to satisfy the requirements of valid publication under the Botanical Code (B 45.4), whereas the names of taxa of organisms not at first considered as animals, if treated as animals are available if they are validly published under the Botanical Code (Z 10 f). For names of taxa of algae, the Botanical Code accepts names available under the Zoological Code, but such a name has to be rejected (B 65) if it becomes a later homonym of a plant name when the taxon to which it is applied is first treated as a plant.

3. The name of a species may not be a tautonym in botany (B 23.4), but tautonyms are permitted in zoology (Z 18, 23 m) at the species rank (e.g., Rattus rattus).

4. The coordinate status of names at family-group, genus-group and species-group levels in zoology (Z 36, 42, 45) is to be contrasted with a restricted autonym rule in botany (B 19, 22, 26). This is a fundamental difference and affects both the dates at which names are considered to have been made available (established) and the operation of priority - coordinate and unaffected by change of rank within the group in zoology (Z 23 c), but generally strictly within a given rank in botany (B 11, 64.3).

5. The concept of illegitimacy in botany (B 63-68), restricting the application of the priority criterion (except for the purposes of homonymy) is not found in zoology, although there are certain analogous restrictions (Z 20, 23 g); likewise, the botanical concept of the superfluous name (B 63) is not embraced by the Zoological Code.

6. Different starting-point dates and works are operative in botany (1753, 1801, 1820, 1840, 1866, 1892, 1900 - B 13) and zoology (1758 - Z 3).

7. Recent (non-fossil) names have priority over fossil names in botany (B 58), except in the case of algae.

8. Later homonyms are to be rejected under both Codes, but the application of the homonymy criterion differs profoundly between the Codes. In botany, the different combinations formed by the same specific epithet with different generic names are considered to be different specific names, while in zoology, the different binomina formed by the allocation of the same specific name to different generic names are considered only as different combinations, not as different specific names. One consequence is the absence in botany of the zoological difference between primary and secondary homonyms (Z 57 b-c); another, that homonyms in botany are identically-spelt names based on different types (B 64); homotypic identically-spelt names are considered to be homonyms in zoology (Z 53 b i, 53 c iii), but in botany are regarded merely as different applications of the same name. The most important difference is that the date of valid publication of a given combination (binomen) has priority for the purposes of homonymy in botany (B 55.1, 56.1), not the date of availability of a given specific name (epithet of botany) as in zoology (Z 52 c). Later specific homonyms are also rejected in botany even if the identically-spelt generic names involved have different types (B 64), whereas in zoology, homonymy of species-group names is to be disregarded if the genus-group names with which the species-group names are combined denote different nominal taxa (Z 57 h i). Lastly, in botany all later homonyms (unless conserved) can and must be rejected (B 64), whereas the zoological concept necessitates provision for the retention or reinstatement of junior secondary homonyms under certain circumstances (Z 59).

9. Form-genera for fossils (B 3.2) and nomina anamorphosium for imperfect states of fungi (B 59) are provided for in botany; collective-group names (e.g., for larval forms) and names of ichnotaxa are analogous in zoology (Z 1 d); but in botany, the earliest name given to the perfect state in pleiomorphic fungi has priority (B 13.6, 59) for the holomorph; no such rule exists in zoology for pleiomorphic animals.

10. The two Codes differ in their concepts of the way in which the type method acts as a link between names and taxa. The Zoological Code regards the nominal taxon as the concept, objectively defined by its type and denoted by an available name, one or more of which may be referred to any taxonomic taxon, whereas the Botanical Code regards the type as the objective criterion common to all applications of a given name. As a consequence, the Zoological Code refers to the type of a nominal taxon (Z 61) while the Botanical Code refers to the type of a name (B 7). Whilst this conceptual difference may by some be regarded as having little practical consequence, the rules to be followed for typification or type fixation also differ in the two Codes (B 7-10 and Guide; Z 61-75); for example, in botany, a misidentified species cannot be made the type of a generic name except by use of the conservation procedure (B 10.3). Also, in botany, the type of the name of a genus is the type of the name of an included species (B 10.1), whereas in zoology, the type of a nominal genus is a nominal species (Z 42 c, 67 a). Living organisms are excluded as types by the Botanical Code (B 9.5) but no such restriction applies in zoology (Z 72 c). The Zoological Code expressly provides for the use of the hapantotype (a suite of preparations of directly related individuals representing different stages in the life cycle) in the case of extant species of protozoa (Z 72 c iv), a provision that has no parallel in the Botanical Code. Finally, the Botanical Code makes the indication of the nomenclatural type a condition for the valid publication of names of new taxa of the rank of a family or below as from 1 January, 1958 inclusive (B 37.1), but such a requirement is not a criterion of availability under the Zoological Code (Z Chapter IV).

11. The rules for orthography and gender of names, and for orthographic correction, differ in the two Codes (B 73, 75; Z 25-32). For example, the standard endings for the names of higher taxa differ (B 16-19; Z 29 a); the decapitalization of specific names (epithets) is mandatory in zoology (Z 28) but sometimes optional in botany (B rec. 73 F); and a demonstrably intentional change in orthography which is not the correction of an incorrect original spelling is treated as not validly published (unavailable) in botany (B 75.1), whereas in zoology such a name is a junior homotypic (objective) synonym having its own author and date (Z 33 b iii). Names ending in -oides are feminine in botany (B Rec. 75 A.4) but masculine in zoology (Z 30 b), while the rules for the use and alteration of the endings -i and -ii, etc., also differ (B 73.10; Z 33 d).

12. Different provisions exist in botany and zoology for the conservation and rejection or suppression of names and works (B 14-15, 69; Z 79-80). Conservation and rejection in botany need not always be absolute, but are subject in some cases to normal operations of priority, while in zoology, names may be totally suppressed or suppressed partially (for purposes of the operation of either homonymy or priority) or conditionally (to confer precedence on one name over another without suppressing either).

13. Rules for the nomenclature of hybrids form part of the Botanical Code (B H 1-12), whereas the nomenclature of hybrids is excluded from the Zoological Code (Z 1 b 3).

It should be clear from the above review that the differences between the Botanical and Zoological Codes are far from insignificant, and that any change in either towards a unified Code would result in so many name-changes in one or other of the disciplines that unification would be less convenient for biologists than continuation of the admittedly imperfect status quo. Codes of nomenclature exist for the convenience of biologists, and since they can function only by voluntary acceptance, acceptance must be more convenient than non-acceptance. The Utopian might imagine an ideal Code, adopting the best provisions of the existing ones, e.g.: effectiveness and validity of publication; correctness of names; coordinate status of names; no illegitimacy or superfluity of names; no autonyms; no nominal taxa; no first reviser principle; combinations and epithets below generic level; priority of combinations in infrageneric homonymy, without primary and secondary homonymy; absolute rejection of all later homonyms; plenary powers for any exceptions, and, as in the International Code of Bacterial Nomenclature, a modern starting-point work and date and an official journal (or journals) for validating names. But in practice, and rapprochement in the two Codes will probably have to be confined to such relatively peripheral matters as the adoption in both of similar terms and definitions, criteria for the (effective) publication of names, and conventions in the citation of authors' names and the use of parentheses.

Acknowledgement

I am grateful to W.D.L. Ride for helpful suggestions.

RESOLUTIONS ADOPTED BY THE IUBS XXIInd GENERAL ASSEMBLY

(1-7 September, 1985, Budapest, Hungary)

WITH RELEVANCE TO TAXONOMY AND BIOLOGICAL NOMENCLATURE

RESOLUTION 3: Scientific Programme

Receiving with satisfaction reports of reviews of the IUBS Scientific Programme committed at the XXIst General Assembly at Ottawa in 1982,

Recognizing a need for a still more extensive programme within the limits of resources available to IUBS and participating bodies and scientists,

Receives and Adopts the Scientific Programme presented to the General Assembly at the XXIInd General Assembly, Budapest.

(except of the accepted report of the Ad Hoc Committee on the IUBS Scientific Programme to the General Assembly)

      3) The International Union of Biological Sciences is urged to appoint an **ad hoc committee on taxonomy** to consider the apparent shortage of trained taxonomists, especially in the Third World countries, and possible measures to alleviate the problem. This committee is charged to
a) obtain data on the amount of basic taxonomic work needed in regard to the fast modification of many ecosystems and associated current biological research.
b) obtain a quantitative estimate on the number of positions available and required for taxonomists, and estimate how many of these have been lost during the last two decades.
c) investigate the type of training that is required to best fill the need, and ways to obtain this training.
The committee is requested to make a first report to the IUBS Executive Committee by 1 July, 1986.

      4) At the suggestion of the section on Zoological Nomenclature, it is proposed that a feasibility study be carried out on the **registration of new scientific names of organisms,** and to include in this project as a pilot project, the desirability of re-publication of the official lists of "Names of Zoology and Works in Zoological Nomenclature" as well as the official "Indexes of Rejected and Invalid Names and Works in Zoology". Special attention should be paid to the financial implication to the Union of the proposed Registry.

RESOLUTION 4: Support for Systems of Nomenclature

Recalling the decisions and conclusions of previous General Assemblies on the fundamental and applied importance of taxonomy, including nomenclature and the need for international support for systems of nomenclature that will secure their continued function,

Appreciating the generous support provided in recent years to Zoological Nomenclature by members of IUBS and, in particular, by the Royal Society (including the British Research Councils), the Australian Academy of Science, the Royal Danish Academy of Sciences and Letters, the South African Council for Scientific and Industrial Research, and the Royal Swedish Academy of Sciences,

Noting with pleasure the donation made by the USSR Academy of Sciences announced at the General Assembly

Commends to all member bodies the importance of developing and providing ongoing support for all systems of biological nomenclature which provides a fundamental base for communication in biological science.

RESOLUTION 5: Response of Systems of Nomenclature to Developments in Science and Information Technology

Noting the report made to the General Assembly by its Ad Hoc Committee on Biological Nomenclature

Noting in particular that the report proposes involvement by the Union in encouraging a harmonious and common approach by the various systems of nomenclature, in facilitating the development of protistan taxonomy and nomenclature, in promoting increased communication between those bodies administering the systems and biologists in general through the International Congresses in Systematic and Evolutionary Biology (ICSEB), and in seeking solutions to difficulties being experienced in botany and zoology as a result of growth of scientific literature and in information technology

Commending the recommendations of the Report to the Executive Committee of IUBS

Requests the Executive Committee to establish a standing committee to report to it at each meeting of the Executive Committee in futherance of the continuing aim of IUBS to achieve maximum harmony between the different systems of nomenclature.

RESOLUTION 6: Importance of Biological Taxonomy

Reiterating its view that proper development of the knowledge of the kinds and distribution of organisms throughout the world is basic to development of biological science

Noting that taxonomic collections are inadequate and poorly supported in many countries, especially in the developing countries

Recalling the Resolution of the XXIst General Assembly in which it noted its pleasure that the Egyptian Academy of Scientific Research and Technology had approved the establishment of a Natural History Museum in Egypt,

Requests the Executive Committee to seek information on projected developments in taxonomy, and

Urges relevant international and national bodies to support development of taxonomy and biological collections.

REPORT OF THE AD HOC COMMITTEE (NOW STANDING COMMITTEE)

ON BIOLOGICAL NOMENCLATURE

1. Meeting

The Committee met in Brighton, England, in July 1985 concurrently with the 3rd International Congress on Systematic and Evolutionary Biology.

As well as meeting in closed session, the Committee attended symposia on « The Protists : Evolution, Taxonomy and Nomenclature » (convened by Dr. J.O. Corliss) and on « Codes of Nomenclature » (convened by Mr. R.V. Melville). For one session of the latter Symposium the members of the **ad hoc** Committee, Dr. W.D.L. Ride (Chairman), Dr. Rita R. Colwell, and Dr. W. Greuter constituted a panel and conducted a free workshop to discuss the future development of various codes of nomenclature.

2. The Committee reports :

(a) Taxonomy and Nomenclature of Protists.

Considering views expressed in formal presentations at the symposium « The Protists : Evolution, Taxonomy and Nomenclature », and the discussions on those presentations, the Committee has concluded that the major issues leading to the present instability in the nomenclature of protists are taxonomic. Protistan taxonomy is currently in a healthy state of active scientific questioning as a result of much new information arising from the application of developing methodologies. The consequence has been the development of fundamentally different classifications at all classificatory levels by different authorities. The Committee CONSIDERS that IUBS should continue to encourage protistologists to meet and exchange information on protistan taxonomy with the aim of reaching common approaches and, if possible, reaching agreement on the composition of major groups.

There is currently consensus that protistan taxonomists will continue to use both the Botanical and Zoological Codes but, since there is no agreement on the composition of major groups, it is not possible to recommend the allocation of different taxonomic categories to particular codes. It is generally agreed that taxonomists will apply the codes of their choice and, in particular, those that are customarily used in determining names in the group concerned. When the use of different codes results in the use of different names for the same taxon, the problem will be solved by the case-by-case approach exemplified by Silva (1980, **Taxon,** vol. 29, pp 121-143).

In order to facilitate the use of the two codes, and to expedite the solution of problems arising from the case-by-case approach, the Committee RECOMMENDS.

(i) that the IUBS Commission on the Nomenclature of Plants (General Committee on Botanical Nomenclature) and the Section on Zoological Nomenclature (International Commission on Zoological Nomenclature) be requested to commence a study of the Codes with the aim of harmonizing them as far as possible, both in essential terminology and in such features as would lead to different names for to the same taxon. In particular, the bodies responsible for the codes are asked to review the appropriateness of introducing.

• provisions to expedite the use of the plenary power of the Commission (zoology) and the procedure for *nomina conservanda*.

• and *nomina rejicienda* (botany) in cases where it is necessary to use such provisions to achieve harmony in the case-by-case approach.

• provisions for the admission of living types and multiple types.

• provisions enabling protistologists to treat protists under one or other of the two codes without being required to define them as either plants or animals.

(ii) that in course of the studies referred to in (i) above, the bodies responsible for the Codes be encouraged to consult whenever appropriate with the following international organizations :

• International Commission for the Nomenclature of Bacteria.
• International Mycological Association.
• International Society for Evolutionary Protistology.
• International Society of Phycologists.
• International Palaeontological Association.
• Society of Protozoologists.

(iii) that a body active in protistan taxonomy (possibly the International Society for Evolutionary Protistology) be encouraged to publish lists of generic names validly published (botany) or available (zoology) for protists.

(b) Consequences to Nomenclature Systems of developments in Information Technology.

68

As a consequence of discussions at the Symposium on the Codes of Nomenclature, the ICSEB III Congress resolved to request the bodies responsible for Botanical and Zoological Nomenclature to establish registers of new names and to make registration a necessary condition for the establishment of a new name in zoology or botany.

The Committee notes that such a procedure is already established in Bacteriological Nomenclature and, because of the increasing inapplicability of requiring conventional publication as a condition of establishing new names, and because of increasing difficulties in learning of newly described and named taxa published in obscure works, RECOMMENDS the endorsement of IUBS sought in the ICSEB Resolution.

(c) Interdisciplinary approach to problems of nomenclature

Considering the success of the two ICSEB symposia in fields of nomenclature in creating a better understanding between biologists working in fields of the taxonomy and nomenclature of viruses, bacteria, plants and animals, the Committee endorses the recommandation of ICSEB III to the Organizing Committee of ICSEB IV to provide a forum at that Congress for continuing the multidisciplinary approach.

(d) Possible function of ICSEB Congresses to provide a venue for official meetings of the IUBS Section of Zoological Nomenclature.

Noting that the functions of the Section of Zoological Nomenclature are to enable zoologists generally to participate with Commissioners of the International Commission on Zoological Nomenclature in electing Commissioners, and in considering proposed amendments to the International Code on Zoological Nomenclature (and recommending on proposed amendments to IUBS), and moreover considering that the attendance of systematic and evolutionary biologists at ICSEB Congresses provides such a forum, the Committee RECOMMENDS that the International Commission on Zoological Nomenclature be requested.

(i) to consider whether such an arrangement would be appropriate and in the interests of zoological nomenclature.

(ii) and, if such an agreement is favoured, to propose for adoption such changes to the Constitution and By-Laws of the Commission and to the Statutes of the Section of Zoological Nomenclature as would be required to enable the arrangement to be implemented.

The attention of the Section on Nomenclature should be drawn to Resolution 1 on the ICSEB III Congress that gives effect to the IUBS Resolution 4 of the XXIst General Assembly that a Group of Systematic and Evolutionary Biology (including the Section of Zoological Nomenclature) be formed within IUBS.

(e) Formation of Advisory Committee to IUBS Executive on interdisciplinary matters in biological nomenclature.

The Committee considers that it would be in the interests of biology for the Executive of IUBS to be informed of developments in different disciplines of biological nomenclature to ensure that the need to achieve maximum harmony is recognised as a continuing aim of IUBS.

Accordingly, the Committee RECOMMENDS that a Standing Committee be established for that purpose to report to the Executive Committee of IUBS.

Resolutions of ICSEB III Referred to in the Above-Mentioned Report

Establishment of Registers of Plant and Animal Names

RECOGNIZING that the growth in scientific literature increases the difficulty of scientists learning of newly described and named taxa of plants and animals and, consequently, the importance of registers of such descriptions and names and of regularly updating these registers;

CONSIDERING that modern developments in publication methods are rendering inoperative the distinction made in Codes of Nomenclature between printed and non-printed material, as a criterion basic to the establishment of new names;

CONSIDERING ALSO that previous difficulties of establishing registration systems for scientific names can now be overcome due to the availability of data-handling and data-management techniques that ensure both central management and decentralized access;

The Congress of Systematic Evolutionary Biologists assembled at ICSEB III

URGES the General Committee for Botanical Nomenclature and the International Commission on Zoological Nomenclature, in association with such other groups with responsibilities in biological nomenclature and information systems as may be interested, to take steps to establish registration procedures for new scientific names in Botany and Zoology;

ENCOURAGES the next International Botanical Congress, through its Nomenclature Section, and the International Commission on Zoological Nomenclature to consider **introducing provisions into their respective Codes to make registration of new names mandatory once satisfactory registration procedures have been established and are operating;**

AND, subject to the agreement of the responsible bodies, REQUESTS the International Union of Biological Sciences (IUBS) to endorse the proposal and to give consideration to means of supporting this commitment of fundamental importance to systematic and evolutionary biology, and to international biological science generally.

Nomenclatural Forum at ICSEB IV

RECOGNIZING the valuable exchange of views amongst the participants in the study of international Codes of Nomenclature at ICSEB III, and the progress made there in handling the nomenclature of organisms, such as protists, that come under more than one Code.

The Third International Congress of Systematic and Evolutionary Biology REQUESTS the Organizing Committee of ICSEB IV to provide a forum at that Congress for continuing multidisciplinary approach to matters relating to the nomenclature of organisms.